CLIMATE-CHANGE HYSTERIA

An Unflattering Demonstration of
Human Gullibility and Ignorance

An Examination of How Demagoguery,
Gullibility, Ignorance, and Mass Media Have
Been Exploited to Perpetrate Global Fraud

"There are none so blind as those who will not
see. The most deluded people are those who
choose to ignore what they already know."
John Heywood (c. 1497 - c. 1580)

"Criticism may not be agreeable, but it is
necessary. It fulfills the same function as
pain in the human body. It calls attention
to an unhealthy state of things."
Winston Leonard Spencer Churchill
(11/30/1874 - 1/24/1965)

DAVID L. R. STEIN

Copyright © 2019 David L. R. Stein
All rights reserved
First Edition

PAGE PUBLISHING, INC.
New York, NY

First originally published by Page Publishing, Inc. 2019

ISBN 978-1-68456-358-6 (Paperback)
ISBN 978-1-68456-359-3 (Digital)

Printed in the United States of America

CONTENTS

The cover photo by Ranga B. Myneni, Professor of Biology at Boston University, can be found at: https://www.nasa.gov/feature/goddard/2016/carbon-dioxide-fertilization-greening-earth. It records the substantial increase in Earth's green "leaf area" over a recent thirty-three-year period and illustrates the well-known fact that increased levels of carbon dioxide in Earth's atmosphere result in increased plant growth which, in turn, results in increased levels of oxygen in Earth's atmosphere. As carbon dioxide is essential to plant life and oxygen is essential to animal life, the cover photo reveals the inconvenient truth that all life on Earth benefits from increased levels of carbon dioxide in Earth's atmosphere. Notwithstanding the outrageous claims by scaremongers that global warming inevitably results in increased desertification, the geophysical record illustrates why climatic episodes associated with high levels of carbon dioxide in Earth's atmosphere are known as "Greenhouse Earth" episodes and why—much to the chagrin of the so-called "Green Movement"—the surest way to "green" the Earth is to increase the amount of carbon dioxide in Earth's atmosphere.

PREFACE

Climate alarmists' continue to issue dire forecasts of global disasters in coming decades based on long-term computer projections of alleged "anthropogenic global warming" (AGW)—supposedly caused by accumulation of atmospheric carbon dioxide produced in the burning of fossil fuels. Examinations of the climate-forecasting models which are producing the dire forecasts reveal them to be primitive—failing to account for the dynamics of all the phenomena known to be forcing Earth's climate changes. At a time when meteorologists' forecasting models can't reliably predict local weather for more than a few days at a time, claims of a capability to reliably forecast Earth's climate for decades or hundreds of years are simply preposterous.

Some observers believe we have been witnessing a classic example of the kind of alarmism immortalized in the fable <u>The Boy Who Cried Wolf</u>[1] in the belief that eventually one of the dire forecasts will prove to be correct. Others are not so kind, for <u>example</u>[2]:

> *"In fact, the <u>IPCC</u>[3]'s new estimate is now identical to Exxon's 1977 estimate and the 1979 estimate of the U.S. National Research Council. In other words, on the crucial question, the help we're getting from climate models has not improved in 40 years and has been going backward of late… But this colossal error not only falsifies the work of the IPCC over the past 28 years, it falsifies the entire climate modeling enterprise of the past half-century."*[1]

---

[1]   *How the Exxon Case Unraveled*, Holman W. Jenkins, Jr., The Wall Street Journal, August 30, 2016

Still others believe it suffices to know that current climate-forecasting models fail to take into account astrophysical and geophysical phenomena known to have dramatic effects on Earth's climate.

> *"CLIMATE IS ONE OF THE MOST COMPLEX SYSTEMS IN NATURE AND IT IS A MAJOR SCIENTIFIC CHALLENGE TO UNDERSTAND THE BASIC MECHANISMS LEADING TO CLIMATE CHANGES. IN THE PAST, CLIMATE SHOWED QUITE REMARKABLE CHANGES OVER ALL TIMES SCALES. In particular, one of the most striking aspects of past climate changes is the so called 100 ky* [thousand-year] *cycles observed in paleoclimatic records. THE CYCLE ITSELF IS NOT AT ALL PERIODIC, SHOWING SUDDEN WARMING PHASES FOLLOWED BY GENTLE DECREASES OF TEMPERATURE. THIS BEHAVIOUR IS IN PHASE WITH THE SO CALLED <u>MILANKOVITCH</u>*[4] *CYCLE,*[2] *I.E. TO THE CHANGE OF GLOBAL RADIATION OF THE SUN DUE TO ASTRONOMICAL CHANGE IN THE EARTH ORBIT...*
>
> *WE LEARN A LOT BY APPLYING STOCHASTIC RESONANCE IN THE THEORY OF CLIMATE CHANGE... IT IS A CRUCIAL STEP TO UNDERSTAND THAT FAST VARIABLES* [with relatively weak effects] *CANNOT SIMPLY BE IGNORED IN THE STUDY OF LONG-TERM CLIMATIC CHANGE...* [*EMPHASES* by author]"[3]

---

[2]   There are three separate Milankovitch cycles, but the author chose the longest one for purposes of demonstration.

[3]   *Stochastic Resonance: from Climate to Biology*, R. Benzi, Nonlinear Processes in Geophysics, Volume 17, Pp. 431-441, 2010. This paper provides a mathematical demonstration of how relatively short-lived and small changes

It's a thesis of this book that the real "climate deniers" are those who are so ignorant of our physical world they don't know the only thing constant about Earth's climate is that it is now, and always has been, constantly changing—and changing so chaotically the changes can be accurately described only as stochastic. Rather than continuing to further inflame climate-change hysteria, isn't it about time to admit the obvious reality that Earth's climate is, and always has been, constantly changing far beyond any feeble human capability to stop it and acknowledge that humanity should focus on accommodating what it can't control?

---

in Earth's solar climatic forcing can be amplified by the internal nonlinear dynamics of stochastic resonances.

INTRODUCTION

What is it about humankind that enables them to cling to erroneous notions in the face of hard realities and contradictory scientific evidence? The short answer is woeful human ignorance combined with a stubborn refusal to learn, an instinctive fear of being ostracized for taking unpopular positions on matters which are perceived to be controversial, and constant bombardment by a steady stream of what commissars of the former Soviet Union accurately termed agitprop[5]—a portmanteau for agitation and propaganda—i.e., disinformation designed to achieve a political agenda.

Unfortunately, a majority of the masses are not now, nor are they ever likely to be, equipped to comprehend many complex geophysical phenomena with geopolitical consequences—never mind their general lack of an ability to critically examine such phenomena. However, their opinions and worldviews are nonetheless strongly shaped by what they hear, read, and see in the mass media. Accordingly, demagogues are able to exploit the gullibility and ignorance of the masses to popularize outrageous falsehoods regarding complex phenomena simply by repeatedly disseminating cunningly crafted agitprop via mass media. As a result, a majority of the masses not only remain ignorant of the true nature of the complex phenomena at issue, but they become indoctrinated in outrageous falsehoods about them as well. All the while—in concerted efforts to prevent exposure of the falsehoods through critical examinations of the agitprop or its sources—its publishers incessantly tout the supposed benefits, integrity, and virtues of a self-policing "free press."

Lest anyone be tempted to believe a self-policing "free press" acts as the virtuous Fourth Estate[6] imagined by Thomas Carlyle[7]— to perform a reliable authentication, fact-checking, and verification service sufficient to prevent mass media from disseminating agit-prop—<u>Ferdinand Lundberg</u>[8] explained why any such notion is a vain pipe dream, to wit:

> *"JOURNALISM, which shapes, modifies, or subtly suggests public attitudes and states of mind, morbidly attracts the owners of the great fortunes, for whose protection against popular disapproval and action there must be a constantly running defense, direct or implied, specific or general. The protective maneuvers often take the form, in this plutocratic* [and subservient] *press, of eloquent editorial assaults upon popular yearnings and ideas. The journalism of the United States, from top to bottom, is the personal affair bought and paid for of the wealthy families* [i.e., by the financially powerful]*... Newspapers all over the world exist, and have existed, in the service of economic and political power rather than in that of truth and noble ideals."*[4]

Merriam-Webster's Unabridged Dictionary defines "climate" as *"the average course or condition of the weather at a particular place over a period of many years as exhibited in absolute extremes, means and frequencies of given departures from these means, of temperature, wind velocity, precipitation and other weather elements."* Note the implicit assumption of constant weather variability.

Earth's climate is known to vary widely with <u>carbon cycles</u>[9], <u>methane cycles</u>[10], <u>Milankovitch cycles</u>[11], <u>solar cycles</u>[12], *<u>El Niño and La Niña</u>*[13] cycles, volcanic eruptions, and other highly variable astrophysical and geophysical phenomena, and no reliable model exists for stochastic perturbations[5] in Earth's absorption of solar

---

[4] *AMERICA'S 60 FAMILIES* by Ferdinand Lundberg, The Vanguard Press, New York, 1937

[5] As no reliable model exists for stochastic perturbations in solar energy output, Earth's albedo, and the coupling of solar energy to Earth's magnetosphere and

energy. How then is it possible for a majority of the world's population to buy into the delusional proposition that the Earth's climate depends only on its atmospheric carbon dioxide content? The claim of AGW scaremongers is that Earth's atmospheric carbon dioxide is now reaching levels that endanger life on earth. As shown in the cover photo and in Figure 1[14], that is a deliberate and outrageous lie promulgated solely for political purposes.

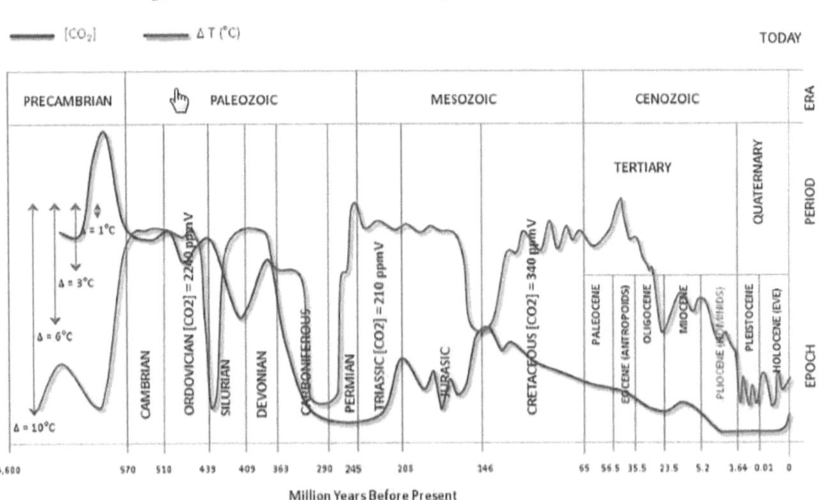

1- *Analysis of the Temperature Oscillations in Geological Eras* by Dr. C. R. Scotese © 2002. 2- Ruddiman, W. F. 2001. *Earth's Climate: past and future.* W. H. Freeman & Sons. New York, NY. 3- Mark Pagani *et all. Marked Decline in Atmospheric Carbon Dioxide Concentrations During the Paleocene.* Science; Vol. 309, No. 5734; pp. 600-603. 22 July 2005. *Conclusion and Interpretation* by Nasif Nahle ©2005, 2007. Corrected on 07 July 2008 (CO2: Ordovician Period).

Figure 1

____

troposphere, current climate-forecasting models are unable to account for known perturbations in Earth's absorption of solar energy.

FACTORS ENABLING MASS DECEPTION AND HYSTERIA

Failing Institutions Corrupted by Political Agendas

Beginning in the nineteenth century—well before creation of the Comintern[15] in March 1919—Neo-Marxist revolutionaries have been gradually, steadily, and stealthily insinuating themselves and their ideology into democratic institutions worldwide. Taking their lead from Lenin, their weapon of choice has been the brain-washing of children and incessant propagandizing of the masses—accomplished through co-optation of educational and governmental institutions and mass media as instruments for disseminating their cunningly crafted agitprop. So successful have they been that:

- Educational institutions have been infiltrated, corrupted, and transformed from sources of objective, open, and rational debate, instruction, and learning into little more than poorly managed unsupervised day-care centers functioning as Neo-Marxist indoctrination camps for expunging all historical, parental, and religious influences while radicalizing students in the guise of education;

- Governmental institutions have been infiltrated, corrupted, and transformed from dedicated civil service organizations into bloated self-appointed, self-aggrandizing, self-perpetuating, self-righteous, self-serving, and unaccountable bureaucracies pursuing their own Neo-Marxist revolutionary agendas; and

- Mass media have been infiltrated, corrupted, and transformed from a potential Fourth Estate into a Fifth Column of Neo-Marxist revolutionary coconspirators hell-bent on overthrowing democratic governments and installing fascist dictatorships in their place.

Along the way, Neo-Marxist revolutionaries have continually relabeled their creed variously as Marxism, Socialism, Communism, Fascism, National Socialism, Progressivism, Modernism, and now <u>Postmodernism</u>[16], etc.—generically <u>Neo-Marxism</u>[17] or <u>Sociocommunism</u>[18] and hereinafter referred to as Neo-Marxism—in ongoing efforts to disassociate themselves from the disastrous and monstrous genocides produced by their many failed attempts to legislate utopia over the last one hundred and fifty years—including responsibility for the unnatural deaths of <u>an estimated one hundred million people</u>[19]accused of being impediments to progress in the last century alone. While the international Neo-Marxist revolutionary campaign has been so insidious that it has largely gone unnoticed by the vast majority of citizens in democratic countries for more than one hundred years, it has finally reached a point where it no longer can be ignored. Indeed, like an abscessed wound ready to burst, the human pain and suffering produced by the international Neo-Marxist revolutionary campaign now calls attention to itself as a pandemic global affliction which demands attention and redress.

Modern Mass Media: Instruments for Inciting the Worst Human Emotions

It was vastly more difficult to perpetrate mass deceptions and propagate mass hysteria among human populations prior to the advent of inexpensive, stealthy, and universally available modern mass media and their perfection as instruments for creating fear and panic by disseminating agitprop—a.k.a., "fake news"—via continuous streams of cunningly crafted disinformation.

Among the earliest convincing demonstrations of the power of mass media as instruments for creating pandemic fear and panic were the 1930s Nazi propaganda newsreels and radio broadcasts and the famous 1938 Orson Welles' "war of the worlds" radio broadcast[20]. Following those demonstrations, it soon became evident that mass media are a demagogue's dream come true, as every demagogue and tyrant in the last one hundred years has had at least an intuitive understanding of what is known to psychologists as the "illusory truth effect[21]"—a.k.a. the "illusion of truth" effect. This effect—which can be, and has been, confirmed by reliable human psychological experiments—is the human tendency to ascribe increased credibility to frequently repeated assertions independently of whether or not they are true. Tom Stafford[22] summarizes the phenomenon as follows[23].

> *"'Repeat a lie often enough and it becomes the truth,'*
> *is a law of propaganda often attributed to the Nazi*
> *Joseph Goebbels[24]. Among psychologists something*
> *like this* [is] *known as the 'illusion of truth' effect...*
> *Our minds are prey to the illusion of truth effect*
> *because our instinct is to use short-cuts in judging*
> *how plausible something is* [presumed to be].*"[6]*

Interestingly, the illusory truth effect is even more effective when combined with a deliberate abuse of language in the form of emotionally charged euphemisms—such as "anthropogenic global warming" instead of "natural climate change"—which have been contrived to exploit the human tendency to gloss over subtle but important semantic distinctions. This fact—which was well known to totalitarian demagogues such as Marx[25], Lenin[26], Hitler[27], and Alinsky[28]—has been cunningly exploited by the so-called "politically correct," or PC, movement—which substitutes Neo-Marxist

---

[6]   *How liars create the 'illusion of truth,'* Tom Stafford, 26 October 2016

censorship and cultural brainwashing for logical argument and rational discourse in an effort to stifle any and all dissenting views.

Today, with universal and relatively inexpensive access to Internet blogs, real-time news feeds, and social networks—not to mention other mass media such as books, magazines, movies, newspapers, radio, and television all available via the Internet—a majority of the world's populations are inundated by a deluge of information and disinformation from modern mass media which would have been inconceivable only twenty-five years ago. Accordingly, it's vastly easier now than it has been at any previous time in human history to quickly create pandemic hysterias among unsuspecting masses of people through nothing more nor less than repeated disseminations of agitprop—including blatant falsehoods and the most outrageous lies—via mass media.

Fortunately for the publishers of modern mass media, their newfound capability to quickly create pandemic hysterias has proven to be hugely profitable. Unfortunately for society, those increased profits and revenues are derived from publication of, and advertisements for, agitprop crafted to incite the worst human emotions and tensions. Evidently, nothing increases mass-media publishers' profits and revenues faster, or more, than inciting the worst human emotions and tensions. This unfortunate reality has created a positive feedback loop in which mass-media publishers first publish agitprop, and then—by publishing a steady drumbeat of reminders—exacerbate whatever emotions and tensions they have stirred up to stimulate pandemic reactions. In so doing, mass-media publishers have, in fact, become demagogues armed with the most effective tool for demagoguery yet conceived.

Thus, we have entered a new era in which every advertiser's and demagogue's dream has been realized, namely: availability of a relatively inexpensive means for shaping public opinion and stampeding the masses—independently of the merits—simply by repeatedly publishing agitprop via mass media. Sadly, this social tragedy has been exploited by demagogues to create and exacerbate a pandemic global climate-change hysteria simply by flooding mass media with their scaremongering polemics.

FACTORS ENABLING PSEUDOSCIENCE

The Marxist Abandonment of Logic

How did it happen that the masses can be so easily panicked into denying reality in favor of believing the unverified claims of pseudoscientists? The roots of this disaster, like those of so many other societal disasters, are directly traceable to the abandonment of logic by Marxists in their on-going attempts to rationalize their cockamamie sophistries. The following seminal analysis[29]. by Ludwig von Mises[30] is at once recognizably correct, totally satisfying, and absolutely devastating.

"It was at this moment that Marx appeared. Adept as he was in Hegelian dialectic—a system easy of abuse by those who seek to dominate thought by arbitrary flights of fancy and metaphysical verbosity—he was not slow in finding a way out of the dilemma in which socialists found themselves. Since Science and Logic had argued against Socialism, it was imperative to devise a system which could be relied on to defend it against such unpalatable criticism. This was the task which Marxism undertook to perform. IT HAD THREE LINES OF PROCEDURE. FIRST, IT DENIED THAT LOGIC IS UNIVERSALLY VALID FOR ALL MAN-

KIND AND FOR ALL AGES. Thought, it stated, was determined by the class of the thinkers; was in fact an 'ideological superstructure' of their class interests. The type of reasoning which had refuted the socialist idea was 'revealed' as 'bourgeois' reasoning, an apology for Capitalism. Secondly, it laid it down that the dialectical development led of necessity to Socialism; that the aim and end of all history was the socialization of the means of production by the expropriation of the expropriators—the negation of negation. Finally, it was ruled that no one should be allowed to put forward, as the Utopians had done, any definite proposals for the construction of the Socialist Promised Land. Since the coming of Socialism was inevitable, Science would best renounce all attempt to determine its nature.

At no point in history has a doctrine found such immediate and complete acceptance as that contained in these three principles of Marxism. The magnitude and persistence of its success is commonly underestimated. This is due to the habit of applying the term Marxist exclusively to formal members of one or other of the self-styled Marxist parties, who are pledged to uphold word for word the doctrines of Marx and Engels as interpreted by their respective sects and to regard such doctrines as the unshakable foundation and ultimate source of all that is known about Society and as constituting the highest standard in political dealings...

LOGIC AND REASONING, WHICH MIGHT SHOW THE ABSURDITY OF SUCH DREAMS OF BLISS AND REVENGE, ARE TO BE THRUST ASIDE. MARXISM IS THUS THE MOST RADICAL OF ALL REACTIONS AGAINST THE REIGN OF SCIENTIFIC THOUGHT OVER LIFE AND ACTION,

ESTABLISHED BY RATIONALISM. IT IS AGAINST LOGIC, AGAINST SCIENCE AND AGAINST THE ACTIVITY OF THOUGHT ITSELF – ITS OUTSTANDING PRINCIPLE IS THE PROHIBITION OF THOUGHT AND INQUIRY, especially as applied to the institutions and workings of a socialist economy...

The Bolshevists persistently tell us that religion is opium for the people. MARXISM IS INDEED OPIUM FOR THOSE WHO MIGHT TAKE TO THINKING AND MUST THEREFORE BE WEANED FROM IT [EMPHASES by author]."[7]

An Ever-Growing Arsenal of Neo-Marxist Sophistries

Widespread misapplications of the pragmatic method of settling metaphysical disputes popularized by <u>William James</u>[31] may have been inspired by knee-jerk reactions to the early twentieth-century realization that classical Newtonian physics was not the complete understanding of the physical universe it had been thought to be and an associated—but unjustified—fear that absolute truth may not be knowable or may not even exist. Whatever its motivation, misapplications of James' pragmatic <u>method</u>[32] as *"a way of interpreting* [unqualified] *ideas by discovering their practical consequences, that is, the difference the truth of the idea would make in human experience"*[8] led some to believe there was no such thing as absolute objective truth, and consequently, they inferred—incorrectly—it was perfectly acceptable to believe whatever would have the most favorable "practical consequences."

---

[7] *DIE GEMEINWIRTSCHAFT*, Ludwig von Mises, from an English translation of the preface to the second German edition, January 1932

[8] *Pragmatism: A New Name for Some Old Ways of Thinking - Summary*, Critical Survey of Literature for Students Ed. Laurence W. Mazzeno. eNotes.com, Inc. 2010 eNotes.com 25 Sep, 2018

Suffice to say, any such inference is an egregious *non sequitur* with no logical basis whatever and a tragic setback for what supposedly is an age of reason. However, once embarked on the slippery slope created by widespread misapplication of the pragmatic method to questions outside the realm of metaphysics, it's no surprise that Neo-Marxists—who already had denied that logic was universally valid—were quick to embrace the new sophistry and add it to their ever-growing arsenal of sophistries for use in proselyting the gullible and ignorant.

Clearly, promotion of such cockamamie sophistries is a catastrophe for all science. Elizabeth Nickson[33] summarizes the current situation as follows[34].

> *"We are finally reaping the rewards of postmodernism…years of radical relativism propagated…in the universities, seemingly unchallenged by parents or university regents adds up to this: People believe that there is no objective truth. Truth has become something to be invented, rather than pursued… If you feel it, only then can it be true."*[9]

Sadly, many young people with the requisite intelligence to know truth is not something to be invented still choose to acquiesce in the postmodernist madness, as they have an instinctive fear of being ostracized for taking unpopular positions on matters which are perceived to be controversial and which have no apparent immediate adverse consequence for their own welfare. Thus, the adolescent and emotionally insecure "go along to get along" crowd make it much easier for the masses of the gullible and ignorant to be misled by postmodernist demagogues, revolutionaries, and seditionists. Of course, this unfortunate moral weakness of callow youth is well known to Neo-Marxists, who exploit it at their every opportunity.

---

[9] *Radical relativism and the war in Iraq*, Elizabeth Nickson, National Post, April 5, 2003

Political Corruption of Scientific Research Driven by Government Funding

President <u>Dwight D. Eisenhower</u>[35] demonstrated extraordinary prescience when, in his farewell address to the nation, he warned the US electorate of the dangers inherent in the increasing dependence of institutional research on government funding, as <u>follows</u>[36].

"In the councils of government, we must guard against the acquisition of unwarranted influence, whether sought or unsought, by the military-industrial complex. The potential for the disastrous rise of misplaced power exists and will persist. We must never let the weight of this combination endanger our liberties or democratic processes. WE SHOULD TAKE NOTHING FOR GRANTED. Only an alert and knowledgeable citizenry can compel the proper meshing of the huge industrial and military machinery of defense with our peaceful methods and goals, so that security and liberty may prosper together. Akin to, and largely responsible for the sweeping changes in our industrial-military posture, has been the technological revolution during recent decades. IN THIS REVOLUTION, RESEARCH HAS BECOME CENTRAL; IT ALSO BECOMES MORE FORMALIZED, COMPLEX, AND COSTLY. A STEADILY INCREASING SHARE IS CONDUCTED FOR, BY, OR AT THE DIRECTION OF, THE FEDERAL GOVERNMENT... THE FREE UNIVERSITY, HISTORICALLY THE FOUNTAINHEAD OF FREE IDEAS AND SCIENTIFIC DISCOVERY, HAS EXPERIENCED A REVOLUTION IN THE CONDUCT OF RESEARCH. PARTLY BECAUSE OF THE HUGE COSTS INVOLVED, A GOVERN-

MENT CONTRACT BECOMES VIRTUALLY A SUBSTITUTE FOR INTELLECTUAL CURIOSITY... THE PROSPECT OF DOMINATION OF THE NATION'S SCHOLARS BY FEDERAL EMPLOYMENT, PROJECT ALLOCATIONS, AND THE POWER OF MONEY IS EVER PRESENT—AND IS GRAVELY TO BE REGARDED. YET, IN HOLDING SCIENTIFIC RESEARCH AND DISCOVERY IN RESPECT, AS WE SHOULD, WE MUST ALSO BE ALERT TO THE EQUAL AND OPPOSITE DANGER THAT PUBLIC POLICY COULD ITSELF BECOME THE CAPTIVE OF A SCIENTIFIC-TECHNOLOGICAL ELITE [EMPHASES by author]."[10]

Those who are familiar with university research are keenly aware of the fact that universities have few, if any, funds of their own for research per se, and consequently, funding for university research projects must come from outside sources. That reality inevitably presents university researchers with the same fund-raising needs as the campaigns of political candidates, and it subjects them to all of the same corrupting pressures and temptations. It would defy common sense and human nature to suggest that university research and researchers are any less corruptible than politics and politicians by the constant pressure to secure sustaining funding. Yet, no government agency monitors university research funding in the same manner as the monitoring of political campaign contributions.

Furthermore, university research projects generally lack the magnitude of funding required by government research programs or government-funded industrial research programs, and consequently, they don't typically require separate funding appropriations which would subject them to governmental and public scrutiny. As a result, the general public is forced to rely upon so-called "peer reviews[37]"—

---

[10] Dwight D. Eisenhower, Farewell Address, 17 January 1961

i.e., reviews of published results by a committee of "peers" liberally salted with associates, colleagues, and friends of the "principal investigators[38]"—of university research to ferret out any potential skullduggery or wrongdoings, and therein lies the danger—something which evidently was known to Eisenhower as indicated by the warning in his 1961 Farewell Address.

It should surprise no one that government funding of industrial and university research has led to the same kinds of collusion and corruption as the "regulatory capture[39]" first popularized by George J. Stigler[40] nearly fifty years ago. Moreover, the collusion and corruption between universities and government is effected in the same way as collusion and corruption between industries and government, namely: by reciprocal rotations of influential officials between them. It was precisely this cozy "I'll scratch your back, if you'll scratch mine" kind of relationship which concerned Eisenhower, who witnessed it firsthand as Supreme Allied Commander in Europe during World War II and later as US President. The major difference in collusion and corruption between industries and government and that between universities and government is in their oversight mechanisms. Whereas the former are subject to governmental and public oversight, the latter are subject only to peer review.

That unfortunate state of affairs is exacerbated whenever university researchers collaborate with researchers in government laboratories, and consequently, find themselves in positions to influence government research funding. Frequent occurrences of such cases have naturally led to the same kind of "I'll scratch your back, if you'll scratch mine" relationships—involving reciprocal rotations of influential officials between universities and government research laboratories—in which each lobbies on behalf of the other. When combined with collusion and corruption between universities and government, the result is a triad of collusion and corruption involving a constant circulation of influential officials between government funding agencies, government research laboratories, and universities.

What possibly could go wrong as a result of such cozy relationships? What indeed but *"domination of the nation's scholars by Federal employment, project allocations, and the power of money"* to quote

Eisenhower. And what could be wrong with that? The answer is that such cozy relationships inevitably produce politicized group think, and—as both politics and group think are antithetical to objective scientific inquiry—it's difficult to imagine anything more destructive of scientific integrity.

Unfortunately, politicized group think is precisely what we are now witnessing in the cabal of government funding agencies, government research laboratories, and universities most of which are engaged in exacerbating the current pandemic global climate-change hysteria. Seizing on the Saul Alinsky[41] tenet popularized by Rahm Emanuel[42] as "*You never want a serious crisis to go to waste*," this triad foresaw the obvious funding advantages of playing along with AGW demagogues like James Hanson[43] and Al Gore[44]. While many climate scientists well know their present climate-forecasting models are far too primitive to make reliable long-term predictions of Earth's climate, evidently some lack the honesty and integrity to admit the complexities of Earth's climate are beyond the scope of their current models—as that would have the undesirable effect of undermining their government funding, personal celebrity, and self-importance.

The Impact of Self-Proclaimed "Elites" on Mass Media and Public Opinion

As has been repeatedly demonstrated throughout human history, neither intelligence, nor knowledge, nor leadership skills are prerequisites for fame or fortune. Yet, in today's world, due to a persistent and unfortunate human propensity to equate fame and fortune with intelligence, knowledge, and leadership, we have produced a resultant class of self-appointed, self-aggrandizing, self-perpetuating, self-righteous, self-serving, and supercilious so-called "elites" presuming to have solutions for every problem real or imagined. These self-proclaimed "elites" more often than not have little or no education or training in logic, mathematics, or science; are heavily indoctrinated with Neo-Marxist pipe dreams; and are aided and abetted by financial support from family fortunes, foreign governments, Neo-Marxist foundations and institutions, and demented billionaires distinguished only by their fame or fortune.

Yet, in vainglorious exhibitions of the blind leading the lame, their disinformation and uneducated opinions regarding astrophysical, climatological, geophysical, meteorological, and other scientific phenomena—about which most of them know precious little or nothing—are frequently paraded before the public and presented as information to be taken seriously. An accidental exposé of this unfortunate reality was recently published in a most unexpected place, viz. in a <u>Scientific American</u>[45] e-book entitled *RETURN TO REASON / The Science of Thought*[46]. The introduction to this marvel of agit-prop entitled "Why Do Facts Fail?" contains the following startling admission by Lisa Pallatroni, the book's editor.

> *"I have a confession to make. Despite the lack of evidence, I am firmly in the anti-Monsanto camp. Working for Scientific American, it's an embarrassing disconnect* [as well it ought to be] *that no amount of data will change a belief I admit is an emotional response to years of reading dystopian fiction… Validity of the source material aside…after I formed an opinion about the risks of certain types of genetic modification, it soon became a belief and difficult to change."*[11]

Objective observers who take the time to read *RETURN TO REASON* likely would say the same about Ms. Pallatroni's belief regarding AGW. Continuing with her explication of why facts fail, Ms. Pallatroni then proceeds to quote from the section entitled *"The Roots of Science Denial"* by "atmospheric scientist" <u>Katharine Hayhoe</u>[47] and Jen Schwartz—a Scientific American journalist—as follows.

---

[11] *RETURN TO REASON: The Science of Thought*, Scientific American, Kindle Edition, 2018

> *"[It's] never the actual science. People have a problem with the implications of science for their worldview and, even more important, for their ideology... THEIR RESULTS ALSO SUGGEST THAT SEEING A BELIEF AS UNTESTABLE INCREASES THE COMMITMENT TO THE ORIGINAL BELIEF [EMPHASIS by author]."*[12]

While both of those assertions by Ms. Hayhoe are undoubtedly true, the latter provides a totally satisfactory explanation for why AGW alarmists and climatologists appear to be so confident of their long-term forecasts. They well know that such forecasts are untestable in their lifetimes, so they'll never be embarrassed or held accountable if those forecasts are ultimately proven to be egregiously in error.

It never occurs to such self-righteous and supercilious victims of politicized group think that the alleged factors which they decry in their "climate denier" and "science denier" strawmen are the bases of their own agendas, biases, and hypocrisies—and are the very hallmarks of so-called "elites." Thus, the <u>dysrationalia</u>[48] attributed to their "climate denier" and "science denier" strawmen are equally applicable to AGW alarmists and climatologists desperately clinging to the their pseudoscience.

Equally appalling, contemptible, and revealing is the reprehensible disparagement of "older white men" which appears in the section entitled "*The Roots of Science Denial*"—an inadvertent indiscretion which reveals a disqualifying and egregious postmodernist political bias which has no place in scientific discourse.

What is vastly more informative and persuasive than the pathetic defense of the AGW hypothesis insinuated into *RETURN TO REASON* is its recognition and acknowledgement of the power of the human confirmation bias, the illusory truth effect, and modern mass media to shape public opinion through frequently repeated assertions independently of whether or not they are true. In the sec-

---

[12] *RETURN TO REASON*, op cit

tion entitled "*Inside the Echo Chamber*" Walter Quattrociocchi summarizes the phenomenon as follows.

> *"By applying the methods of computational social science to the traces that people leave on Facebook, Twitter, YouTube and other such outlets, scientists can study the spread of conspiracy theories in great detail. Thanks to these studies, we know that humans are not, as has long been assumed, rational. Presented with unfiltered information, people will appropriate that which conforms to their own thinking. This effect, known as confirmation bias, fuels the spread of demonstrably false arguments—theories about global mega conspiracies, connections between vaccines and autism, and other nonsense. And unfortunately, there seems to be no easy way to break this cycle."*[13]

Ironically, there is no better, nor more conspicuous, example of that phenomenon than the mass hysteria surrounding the AGW hypothesis. What is missing from the polemics in *RETURN TO REASON* is acknowledgement of the fact that self-appointed, self-aggrandizing, self-perpetuating, self-righteous, self-serving, and supercilious so-called "elites" are not only not immune to this phenomenon, they are—as demonstrated by the agitprop and pseudoscience constantly being peddled in Scientific American—among the greatest victims of dysrationalia. As self-exalting "keepers of the faith," these self-proclaimed "elites" exude such an all-knowing confidence in their superiority over mere mortals, they don't hesitate to pontificate about matters about which they are largely, if not completely, ignorant, and—due to their greater mass-media presence and notoriety—they get vastly more media coverage and exposure than more competent and more knowledgeable information sources. As unreasonable as the AGW agitprop insinuated into *RETURN TO REASON* is when it's examined objectively, evidently

---

[13] *RETURN TO REASON*, ibid

it wasn't enough for the truly AGW religious, and—sensing a profit opportunity—Scientific American quickly followed its publication of *RETURN TO REASON* with publication of *CLIMATE CHANGE / Planet Under Pressure*[49]. As an indication of the absurdities to be found in this collection of its previously published polemic screeds, the section entitled "*The Great Climate Experiment*" begins with the following asinine statement.

> *"Business, technology and government forecasts usually look five or 10 years out, 50 years at most. Among climate scientists, there is some talk of century's end. In reality, carbon dioxide dumped into the atmosphere today will affect Earth hundreds of thousands of years hence."*[14]

Unless completely blinded and gripped by pathological fear, only a troglodyte could believe such drivel. Current climate science hasn't the slightest possibility of being able to forecast the carbon dioxide component of Earth's atmosphere that far into the future. To claim otherwise without first being able to simulate the carbon and methane cycles responsible for "Greenhouse Earth" and "Icehouse Earth" episodes[50] is so incredible it's simply preposterous. The author then goes on to state the following.

> *"One of the greatest uncertainties in climate prediction is the amount of CO_2 that will ultimately be released into the atmosphere."*[15]

Clearly, the obvious hypocrisy and irony implicit in that admission escaped the author. While true, it says nothing about the Earth's difficult-to-measure and highly complex and dynamic carbon cycle. The data needed to predict the net effects of Earth's carbon cycle for

---

[14] *CLIMATE CHANGE: Planet Under Pressure*, Scientific American, Kindle Edition, 2018
[15] *CLIMATE CHANGE*, op cit

periods of more than a few decades simply don't exist. Consequently, it's not possible to test any hypothetical model of the carbon dioxide component of Earth's atmosphere over longer periods of time—even in the absence of any powerful influences by Earth's unpredictable volcanoes.

Sadly, more perfect examples of "the pot calling the kettle black" could not be found than in the polemic diatribes regarding climate pseudoscience contained in *RETURN TO REASON* and *CLIMATE CHANGE*. As incongruous as it is for a publication claiming to popularize scientific knowledge, long-time subscribers have come to expect political polemics and posturing in the pages of the increasingly unscientific Scientific American. Yet, the fact that ideologically driven slanders of others' agendas, biases, and hypocrisies by its authors, editors, and referenced "experts" receive no disqualification or reproof by the publisher provides eloquent testimony to Scientific American's failure to adhere to any meaningful ethical, journalistic, or scientific standards. Accordingly, the indictments of others' agendas, biases, and hypocrisies by its authors, editors, and referenced "experts" are revealed to be smokescreens designed to conceal their own agendas, biases, and hypocrisies. Long-overdue and well-contemplated introspection surely could prove to be enlightening, humbling, and rewarding experiences for such self-righteous and supercilious victims of politicized group think.

There was a time in the US when weekly and monthly periodicals still had a sense of journalistic integrity and were a cut above the "fish-wrap" quality of daily newspapers. Tragically, that's no longer the case. Evidently, the extensive, coercive, and corrupting Neo-Marxist brainwashing throughout US educational and governmental institutions and mass media—combined with the irresistible appeal of publishing political agitprop to enhance profits and revenues—simply overwhelmed any pretense of journalistic honesty, integrity, and objectivity. Sadly, most US periodicals now fall into the "fish-wrap" category—having been coerced and corrupted by the same pernicious factors as daily newspapers and other mass media. One gets an appreciation for the horrific social costs of this disaster when one realizes that even what is reported in publications like Scientific American is more agitprop than science.

FACTORS ENABLING THE NEO-MARXIST CONSPIRACY

Also-Rans Seeking to Punish the US for Their Own Relative Failures

The list of parties to the Paris Agreement[51] includes all the countries which are jealous of US achievements, economic successes, and exceptionalism—i.e., every other country in the world. It matters not to the citizens of also-ran countries that it was the US which saved the world from itself in two world wars in the last one hundred years and may be called upon to do so again, that it was the US which stymied world domination by the former Soviet Union, or that it was the US which has been repeatedly called upon to stop totalitarian genocides in countries around the world. It's just too difficult for them to acknowledge that their countries' failures to achieve the greatness of the US is no one's fault but their own. The Paris Agreement is a perfect demonstration of why von Mises was correct when he so cogently articulated the single most-important reason for Marxism's persistent refusal to die—in spite of its unbroken record of failed states and monstrous human carnage—as follows.

> *The incomparable success of Marxism is due to the prospect it offers of fulfilling those dream-aspirations and dreams of vengeance which have been so deeply embedded in the human soul from time immemorial. It promises a Paradise on earth, a Land*

of Heart's Desire full of happiness and enjoyment, and—sweeter still to the losers in life's game— humiliation of all who are stronger and better than the multitude."[16]

Even a deadly pandemic global plague could not serve the purposes of Neo-Marxist demagogues better than pandemic global climate-change hysteria—which, of course, is perceived by the masses to be a slowly evolving equivalent. While other natural disasters—such as asteroid impacts, earthquakes, hurricanes, tsunamis, typhoons, and volcanic eruptions—are terrifying in their own right, even supervolcanoes can't incite as much apocalyptic fear and panic as simultaneously experiencing increasingly violent storms, insufferable heat, melting icecaps, rising seas, spreading deserts, and worldwide crop failures—all portrayed as inevitable consequences of AGW and all leading to a future armageddon. Thus, pandemic global climate-change hysteria is a perfect vehicle for frightening all the world's populations into rising up and demanding actions be taken to prevent AGW. And, as any such undertaking begs the questions of who is going to manage it and who is going to pay for it, pandemic global climate-change hysteria is an ideal stalking-horse for globalist demagogues seeking to establish Neo-Marxist world government.

Give the Neo-Marxists credit for recognizing and seizing this unprecedented opportunity to implement their century-old dream of world conquest. By exploiting pent-up global jealousies of US achievements, economic successes, and exceptionalism; condemning the US as a primary causal agent of AGW and a wealthy exploiter of the less fortunate; and exploiting farcical and inept international organizations—principally the United Nations and its so-called International Court of Justice—to coerce the US into paying reparations to all who can concoct a grievance caused by AGW; Neo-Marxist demagogues hope to bring the US to heel and rob its treasury to implement their dream of "leveling the playing field" by establishing a Neo-Marxist world government. The net effect of this conspir-

---

[16] Ludwig von Mises, op cit

acy would be to reduce the world economy to its lowest common denominator, and—if history is any guide—to ultimately produce a global economic and social regression vastly more devastating than that experienced during the <u>Dark Ages</u>[52] in Western Europe—the notorious birthplace of Neo-Marxism.

Demagogues Seeking to Maximize Fame, Fortune, and Political Power

The insidious Neo-Marxist conspiracy to establish a totalitarian world government has succeeded in recruiting both unwitting and witting dupes from all around the world. Lest any believe the US is unaffected by the conspiracy, the capitulation of the Obama administration to the Paris Agreement provides convincing evidence that a majority of the US Democrat party is complicit in it. How could this happen in the US?

Tragically, with the advanced stage of Neo-Marxist politicized corruption evident in the vast majority of the world's educational and governmental institutions, it's hardly surprising the US is not immune to this plague. What is worse is that a small number of demented, egocentric, gullible, ignorant, misguided, self-aggrandizing, self-righteous, self-serving, and supercilious billionaires persist in attempting to reengineer human society to conform to their own personal fantasies of a better world. As for why this should be the case, <u>John Dalberg-Acton</u>[53]—a.k.a., Lord Acton—probably said it best when he declaimed as follows.

> *"Power tends to corrupt, and absolute power corrupts absolutely. Great* [i.e., powerful] *men are almost always bad men, even when they exercise influence and not authority..."*[17]

---

[17] Letter to Mandell Creighton from Lord Acton, April 1887

As the US has the world's largest economy, it's not surprising it has its share of such luminaries, some of whom—e.g., Jeff Bezos[54], Michael Bloomberg[55], Thomas Steyer[56], S. Donald Sussman[57], and George Soros[58]—are evidently practicing Neo-Marxists, and all of whom—except for their evident egocentricity, gullibility, and ignorance—should know better. To make matters worse, perhaps due to its unique freedoms and liberties—none of which would be long tolerated in the "better worlds" they envision—the US has too many such Neo-Marxist political demagogues to list here.

However, one of them, viz. Al Gore, deserves special dishonorable mention in connection with AGW, as he was the Judas goat[59] who was responsible for inciting fear and panic in millions of the gullible and ignorant—transforming AGW into a new kind of religion. Indeed, Mr. Gore's despicable performance is a quintessential example of just how easy it is for demagogues to exploit modern mass media to incite pandemic global hysterias—based on outright falsehoods supported by disinformation—for their own personal profit. What is especially revealing and troubling about such demagogues and luminaries expounding on subjects of which they are largely, if not completely, ignorant is knowing theirs is a political agenda, and facts, logic, and real science are of no interest to them. In the "ends justify the means" mentality of Neo-Marxists, all that matters is achieving their cherished political goals.

Mercenaries Seeking to Maximize Foreign Markets, Profits, and Revenues

As the twenty-first-century cybernetic business model is global in scope, US-based mass media, multinational corporations, manufacturers of consumer products, professional sports franchises, and purveyors of cybernetic media are among the most zealous globalists—all pandering to foreign audiences and clientele and prostituting themselves to maximize their foreign markets, profits, and revenues.

Sadly, as the majority of their audiences and clientele reside outside of the US, many of these zealots have no compunction whatever about cloaking themselves in globalist Neo-Marxist imagery antithetical to the US Constitutional principles and founding values which are the cornerstones of the US exceptionalism which enabled their success and facilitated their accumulation of power and wealth.

While international business practices tend to be homogenized through global commerce, common international business practices do nothing to establish a common global culture—which would be a prerequisite to establishment of a stable democratic sovereign world government. Of course, Neo-Marxists have no intention of establishing any kind of democratic government. Theirs is a totalitarian model of empire like that of the former Soviet Union and the People's Republic of China.

Having learned nothing from the collapse and failure of the former Soviet Union—or from the fact that no empire which ever attempted to span a large number of different cultures has ever succeeded in bridging the chasms which separate them—despite the fact that most bankrupted themselves in failed attempts to do so, Neo-Marxists are hell bent on eclipsing the failure of the former Soviet Union with one of truly global scale. Tragically, they are being aided and abetted in their idiotic venture by unwitting and witting dupes in the US business community who are willing to maximize their foreign markets, profits, and revenues whatever the costs may be to US achievements, economic successes, and exceptionalism—not to mention US sovereignty.

While the Neo-Marxist lust for global dominion and supreme political power—and the unwitting and witting collusion of demagogues—have been part of the geopolitical landscape for more than one hundred years, advent of cybernetic media, the explosive growth of profits and revenues from foreign sources, and a consequent proliferation of US globalist mercenaries have changed the geopolitical landscape in recent decades. Tellingly, US President Donald J. Trump[60] is receiving stiff opposition from such misguided globalist mercenaries as he attempts to ensure future US achievements, economic successes, and exceptionalism by enforcing US sovereignty over its geopolitical interactions and trade with foreign countries.

A Brave New World

Given the corrosive and corrupting influence of money on institutional research, mass media, and politics, it should surprise no one that the corrosive and corrupting influence of money on all three has the same effect on political malfeasance and skullduggery as an accelerant has on fire. Thus, it's no accident pandemic global climate-change hysteria spread as quickly as it did.

An unanswered question is whether or not the funding of institutional research, mass media, and political demagogues by demented billionaires, foreign governments, and mercenary businesses—with no fidelity to US Constitutional principles and founding values—will ensure the success of the global Neo-Marxist conspiracy. Only time will tell, but one thing is certain: only the US has the power to stop it.

Sadly, recent US history is not reassuring, as the US government is still struggling to recover from years of abuses of modern mass media by its own intelligence, law-enforcement, and security agencies. Unfortunately, such abuses are so difficult to detect, prevent, and stop, it can take months or even years to completely unravel them after they've been detected.

Consequently, the new rabble-rousing capabilities of modern mass media have enabled an entirely new kind of asymmetric cybernetic guerilla warfare, and it remains to be seen whether or not US intelligence, law-enforcement, and security agencies can recover from their own abuses of these same capabilities to preserve and protect US sovereignty by mounting an effective defense against them.

Unfortunately, with the aid of modern mass media, the unique US freedoms of assembly and speech are now being effectively turned against it by foreign enemies, as well as by domestic revolutionaries and seditionaries. In addition to providing such subversive forces with an inexpensive, stealthy, and universally available means to disseminate their agitprop—frequently while escaping detection—the same mass-media tools are being exploited to quickly incite large numbers of the gullible and ignorant for any number of nefarious purposes—including, but not limited to, organizing <u>flash mobs</u>[61] or massive public demonstrations to create civil disobedience, disrupt civil proceedings and public events, impede law enforcement, incite violence, intimidate adversaries, and protest policies unfavorable to their cause.

The reprehensible and subversive "Antifa[62]" demonstrations now taking place in cities across the US are perfect examples of how modern mass media are being employed to exploit the unique US freedoms of assembly and speech for seditionary purposes. Such despicable "protests" have nothing whatever to do with "government of the people, by the people, for the people" as so eloquently articulated by Abraham Lincoln[63] at Gettysburg[64]. Instead, they are exhibitions of totalitarian thuggery. Funded by demented billionaires and foreign governments, they are reminiscent of the "storm trooper" rallies of the Nazi *Sturmabteilung*[65] which operated in the early days of the Third Reich[66]. When modern mass media—with as many applications for evil as for good—are employed to facilitate such dangerous seditions—fueled by large numbers of otherwise unemployable or unemployed people paid by, and acting as stooges for, demented billionaires or foreign governments—they are directly or indirectly responsible for creating an explosive revolutionary and seditionary cocktail which threatens continuation of civil and constitutional government.

Modern societies will require new laws and regulations to protect them from being ravaged by the new kind of asymmetric cybernetic guerilla warfare facilitated by modern mass media and waged by anarchists, political demagogues, revolutionaries, seditionaries, and terrorists. Unfortunately, social media are perfect examples of why it takes far too long for laws and regulations to catch up to new technologies. Uses and abuses of social media have been rampant for more than a decade, yet social media continue to be totally unregulated.

The time is long past for the US to face up to its profligacies and over-tolerance of revolutionary and seditious forces which have been operating within its borders for more than one hundred years. If it cannot rouse itself to moderate those excesses, the ultimate demise of the US as constituted by its founders will stand as eloquent testimony to the horrible truth of <u>Henning W. Prentis, Jr.</u>[67]'s assessment of humanity's seeming inability to learn from its past mistakes and prevent recurrent backsliding from periods of social progress into periods of social regression, which he articulated as <u>follows</u>[68].

> *"Paradoxically enough, the release of initiative and enterprise made possible by popular self-government ultimately generates disintegrating forces from within. Again and again after freedom has brought opportunity and some degree of plenty, the competent become selfish, luxury-loving, and complacent; the incompetent and the unfortunate grow envious and covetous, and all three groups turn aside from the hard road of freedom to worship the Golden Calf of economic security. The historical cycle seems to be: from bondage to spiritual faith; from spiritual faith to courage; from courage to liberty; from liberty to abundance, from abundance to selfishness; from selfishness to apathy, from apathy to dependency; and from dependency back to bondage once more."*[18]

Interestingly, the present US plight has not gone unnoticed by astute observers. For example, <u>Kenneth Archer</u>[69], a self-styled "*Well read, traveled, synergistic, skilled, educator, philosopher, crazy old man,*" made the following <u>cogent observation</u>[70] in response to the question: "How hard would it be to conquer the US?"

> *"It depends on what you mean by 'conquer.' Conquer as in a war; it would be difficult. Conquer does not require violence. The US is already showing signs of being conquered. The US is being conquered from within, and democracy is accomplishing it with great speed. The biggest downfall of a democracy is*

---

[18] *The Cult of Competency*, by Henning W. Prentis, Jr., Mid-Year Convocation of the General Alumni Society of the University of Pennsylvania, 1943

when the people discover that they can vote them-selves luxuries and a better lifestyle but agree not to fund it with taxation. The US is failing because of its debt. How would the banking industry survive if Americans can raise their debt limit on their credit cards with no intentions to repay it?

In the name of equality, political correctness, self-entitlements, self-righteousness, exceptional superiority, nobody is responsible for themselves and an unknown 'they' are responsible for everything. No one wants to do hard manual labor, so machines are invented to do the hard work, yet we think we should earn a living wage and be cared for when doing nothing."[19]

The US can be conquered only from within, and cybernetic guerilla warfare makes that more likely than ever before. Concerned citizens had better do more than sit idly by or their fate is sealed.

"The tree of liberty must be refreshed from time to time with the blood of patriots and tyrants."[20]— <u>Thomas Jefferson</u>[71]

---

[19] *"How hard would it be to conquer the US?"* Kenneth Archer, May 17, 2017

[20] *JEFFERSON QUOTES & FAMILY LETTERS,* "Extract from Thomas Jefferson to William Stephens Smith," Thomas Jefferson Foundation

INCONVENIENT FACTS

The Scientific Method

As basic, simple, and straightforward as it is, evidently the scientific method is not well understood by some who profess to be scientists. For example, in the section of *RETURN TO REASON* entitled "*Cognitive Ability and Vulnerability to Fake News*," the authors state that:

> "*At a more general level, this research underscores the threat that fake news poses to democratic society. The aim of using fake news as propaganda is to make people think and behave in ways they wouldn't otherwise—for example, hold a view that is contradicted by overwhelming scientific consensus.*"[21]

What is alarming about that statement is the phrase "*contradicted by overwhelming scientific consensus*"—which is an obvious oxymoron. Consensus is a political process, whereas scientific facts or truths are derived only through <u>the scientific method of research</u>[72], which has nothing whatever to do with consensus or politics. Only an imbecilic true believer of the cockamamie sophistry known as postmodernism—which teaches that truth is "something to be invented, rather than pursued"—could possibly believe otherwise.

---

[21] *RETURN TO REASON*, ibid

As everyone who is familiar with the scientific method well knows, scientific hypotheses can't be confirmed or tested by consensus or, what is the same thing, by popularity polls. And, as recognized by many of the authors of *RETURN TO REASON*, consensus and politics are easily manipulated via mass media due to the power of the human confirmation bias and the illusory truth effect. Accordingly, the expression "overwhelming scientific consensus" is equivalent to "overwhelming scientific ignorance"—a truly alarming indictment of anyone claiming to be a scientist.

Albert Einstein[73]'s general theory of relativity[74] was not confirmed by consensus. At the outset there was only one person who believed his theory to be true, viz. Einstein himself. That is typical of science, as most scientific breakthroughs result from the efforts of one or a very few persons.

Ironically, the mistaken notion that the scientific method is a consensus process is a result of massive postmodernist brainwashing of the ignorant by the agitprop known as fake news and discussed at length in *RETURN TO REASON*. Scientific hypotheses are confirmed only by unequivocal and verifiable scientific observations of predicted phenomena which can't be otherwise explained. Either scientific hypotheses stand up to every such test, or they must be rejected. Anything else is pseudoscience or, at best, untested speculation and theorizing.

The AGW hypothesis is an example of the latter. Thus, it must be rejected, at least until such time as it has been tested[75] in a manner similar to that of Einstein's general theory of relativity—i.e., via unequivocal and verifiable scientific observations confirming predicted phenomena which can't be otherwise explained. The AGW hypothesis flies in the face of the only real climate science we have today and has yet to pass any such test.

Earth's Chaotic Climate History

In its geologic history, Earth has been covered by both sheets of ice and tropical rain forests, and—as shown in Figure 2—in just the last one thousand years the Earth has experienced both a <u>Medieval Warm Period and a Little Ice Age</u>[76]. Moreover, no reliable model currently exists for stochastic perturbations in either solar energy output or its coupling to Earth's climate, yet it's been known for centuries that Earth's climate and temperatures are highly correlated with solar activity.

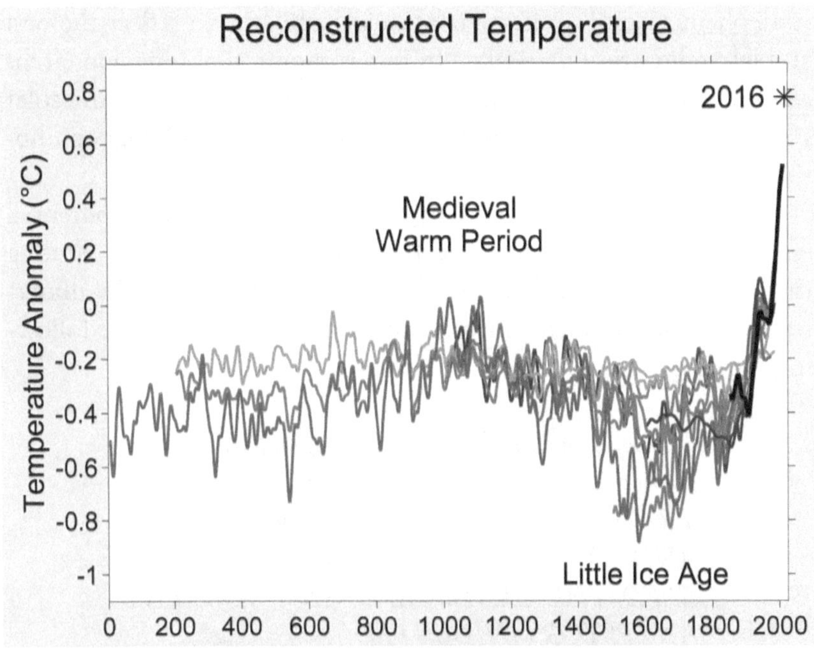

Figure 2

Earth's Chaotic Astrophysical Environment

The solar system has been metaphorically described as a giant "<u>cosmic shooting gallery</u>[77]" with millions of <u>asteroids</u>[78] and <u>comets</u>[79] traveling through space in orbits around the Sun. The erratic orbits of some of the smaller objects cause them to occasionally cross the orbits of other asteroids, comets, and <u>planets</u>[80] where they can have close encounters which can perturb the orbits of both objects, or they can have violent collisions which can completely destroy the smaller objects while changing the orbits of the larger objects. Interestingly, this chaotic dance is not limited to asteroids and comets. While following more stable orbits, the planets also participate in this dance through their own gravitational interactions and their unique orbital dynamics. Nothing orbing the Sun in this cosmic shooting gallery has an unchanging orbit. Consequently, astronomical ephemerides[81] must be continually updated to reflect the best and latest empirical data taken from astronomical observations. This chaotic dance is a far cry from the oversimplified view of the solar system as a system which has reached a kind of circulating equilibrium or stasis.

Some of Earth's orbital perturbations[82] and their potential effects on Earth's climate have been studied for decades. Among those which have been extensively studied, yet are not fully understood, are the so-called Milankovitch cycles described in the following excerpts[83].

> *"Changes in Earth's orbit have helped pace climatic change for millennia. SCIENTISTS ARE NOW TRYING TO UNDERSTAND WHETHER – AND HOW – THESE CHANGES REMODELED THE LANDSCAPES OUR ANCIENT ANCESTORS INHABITED...*
>
> *The pattern of incident solar radiation (insolation[84]) received on the planet at a given place and time is an important factor in understanding both directional trends and variability observed in many paleoclimatic records, particularly those related to Quaternary ice ages... Changes in insolation are, in turn, driven by Earth's natural orbital oscillations, termed Milankovitch cycles. THE THREE ELEMENTS OF MILANKOVITCH CYCLES ARE ECCENTRICITY [Earth's orbital shape, with a period of ~100 KY and a beat[85] period of ~413 KY], OBLIQUITY [Earth's axial tilt from a*

normal to the ecliptic, with a period of ~41 KY], *AND PRECESSION* [Earth's spin-axis orientation, with a period of ~19-23 KY]... *Although the interactions between orbital parameters are major external drivers of paleoclimatic changes, the internal dynamics of the climate system also exert important controls on temporal and spatial patterns of environmental change. Furthermore, BOTH EXTERNAL AND INTERNAL FORCING MECHANISMS CAN INVOLVE A COMPLEX SERIES OF FEEDBACKS, AND RESPONSES THAT MAY BE LINEAR OR NONLINEAR, SYNCHRONOUS OR DELAYED, OR HAVE A CRITICAL THRESHOLD ('TIPPING') POINT* [*EMPHASES* by author]."[22]

These quasiperiodic phenomena have significant stochastic perturbations and, as noted in the following excerpt[86], poorly understood couplings to Earth's climate.

"Artifacts show that the variation in Earth's climate is much more extreme than the variation in the intensity of solar radiation calculated as the Earth's orbit evolves. IF ORBITAL FORCING CAUSES CLIMATE CHANGE, SCIENCE NEEDS TO EXPLAIN WHY THE OBSERVED EFFECT IS AMPLIFIED COMPARED TO THE THEORETICAL EFFECT [*EMPHASIS* by author]."[23]

---

[22] *Milankovitch Cycles, Paleoclimatic Change, and Hominin Evolution*, Christopher J. Campisano, Institute of Human Origins, School of Human Evolution & Social Change, Arizona State University) © 2012 Nature Education

[23] *Milankovitch cycles, Effect exceeds cause*

Interestingly, as noted in the following excerpt[87], Milankovitch cycles are highly correlated with the methane component of Earth's atmosphere, and hence with Earth's global methane cycles.

"SPECTRAL ANALYSES OF THE VOSTOK CH$_4$ [i.e. methane] AND TEMPERATURE RECORDS REVEAL THE PRESENCE OF ALL MAJOR MILANKOVITCH [quasi] PERIODIC-ITIES, namely approximately 100, 41, 23, and 19 KYR. THESE FINDINGS ILLUSTRATE THE IMPORTANT ROLE OF METHANE IN THE PAST, PRESENT, AND FUTURE GREEN-HOUSE EFFECT [EMPHASES by author]."[24]

Earth has other orbital perturbations which have received less scrutiny, including: apsidal precession[88] of Earth's orbit around the Sun—with a period of ~23 KY, and Earth's orbital inclination[89] to the ecliptic—with a period of ~70 KY. Suffice to say, climatologists have yet to measure and model the effects of its orbital perturbations on Earth's insolation—not to mention potential stochastic resonances[90] with unrelated changes in Earth's climate.

Earth's Chaotic Geophysics

Earth's geophysical phenomena appear to be relatively stable over time spans of a few years or even a few decades, but they have been anything but stable over longer time periods. A number of poorly understood phenomena contribute to the chaos. As indicated in the following comments[91] by Patrick J. Michaels[92], a climatologist and director of the Center for the Study of Science at the Cato Institute, it's long been known that the Southern Pacific Ocean phe-

---

24 *The Global Methane Cycle*, Martin Wahlen, Annual Review Of Earth And Planetary Sciences, Annual Reviews Inc., 1993

nomena known as *El Niño* and *La Niña* have <u>direct effects on global weather</u>[93].

> "*Instead of relying on debatable* [inconsistent and error-prone] *surface-temperature information* [as is done by the IPCC], *consider instead readings in the free atmosphere (technically, the lower troposphere) taken by two independent sensors: satellite sounders and weather balloons. As has been shown repeatedly by University of Alabama climate scientist John Christy, since late 1978 (when the satellite record begins), THE RATE OF WARMING IN THE SATELLITE-SENSED DATA IS BARELY A THIRD OF WHAT IT WAS SUPPOSED TO HAVE BEEN, ACCORDING TO THE LARGE FAMILY OF GLOBAL CLIMATE MODELS NOW IN EXISTENCE. Balloon data, averaged over the four extant data sets, shows the same.*
>
> *IT IS THEREFORE PROBABLY PRUDENT TO CUT BY 50% THE MODELED TEMPERATURE FORECASTS FOR THE REST OF THIS CENTURY. DOING SO WOULD MEAN THAT THE WORLD – WITHOUT ANY POLITICAL EFFORT AT ALL – WON'T WARM BY THE DREADED 3.6 DEGREES FAHRENHEIT BY 2100 THAT THE UNITED NATIONS REGARDS AS THE CLIMATE APOCALYPSE.*
>
> *THE NOTION THAT WORLDWIDE WEATHER IS BECOMING MORE EXTREME IS JUST THAT: A NOTION, OR A TESTABLE HYPOTHESIS. AS DATA FROM THE WORLD'S BIGGEST REINSURER, MUNICH RE, AND UNIVERSITY OF COLORADO ENVIRONMENTAL-STUDIES PROFESSOR ROGER PIELKE JR. HAVE*

SHOWN, WEATHER-RELATED LOSSES HAVEN'T INCREASED AT ALL OVER THE PAST QUARTER-CENTURY. IN FACT, THE TREND, WHILE NOT STATISTICALLY SIGNIFICANT, IS DOWNWARD. Last year [2015] showed the second-smallest weather-related loss of Global World Productivity, or GWP, in the entire record. WITHOUT EL NIÑO, TEMPERATURES IN 2015 WOULD HAVE BEEN TYPICAL OF THE POST-1998 REGIME [EMPHASES by author]."[25]

In addition to the phenomena known as *El Niño* and *La Niña*, the Earth is known to experience less frequent—and less predictable—volcanic eruptions of varying durations and power, as well as changes to the geographical orientation of its magnetic poles. As supervolcanoes[94], fluctuations in Earth's magnetic field[95], and geomagnetic reversals[96] have not occurred in recorded human history, their effects on Earth's climate are highly speculative. Earth's volcanic history[97] and prediction challenges are particularly problematic for climatologists, as discussed in the following excerpts[98] from *"Volcanic Activity: Frontiers and Challenges in Forecasting, Prediction and Risk Assessment."*[26]

"MANY VOLCANIC PROCESSES ARE STOCHASTIC [i.e. probabilistic], AND NEED TO BE CHARACTERIZED BY STATISTICAL MODELS. COMPLEX NONLINEAR SYSTEMS CAN BE VERY UNSTABLE CLOSE TO

---

[t] *The Climate Snow Job*, Patrick J. Michaels, The Wall Street Journal, January 24, 2016

[26] *THE STATE OF THE PLANET: FRONTIERS AND CHALLENGES IN GEOPHYSICS, "Volcanic Activity: Frontiers and Challenges in Forecasting, Prediction and Risk Assessment,"* R. S. J. Sparks and W. P. Aspinall, the International Union of Geodesy and Geophysics and the American Geophysical Union, 2004

CRITICAL THRESHOLDS BETWEEN STABLE STATES, BECOMING INHERENTLY UNPREDICTABLE…

In volcanology numerical modelling is becoming an important aspect of forecasting and risk assessment, and evolving approaches have largely focused on numerical simulations. Aided by increasing computer power as well as improving understanding of the physics involved, such models are becoming increasingly sophisticated. A PRESUMPTION IS THAT SUCH MODELS WILL EVENTUALLY SIMULATE NATURE SO WELL THAT THEY CAN BE USED FOR FORECASTING. THESE EXPECTATIONS MAY PROVE TO BE OPTIMISTIC, NOT LEAST BECAUSE EXPERIENCE HAS DEMONSTRATED WITH CLIMATE MODELLING THAT, WHEN UNCERTAINTIES IN SUCH MODELS ARE DISAGGREGATED AND APPRAISED INDIVIDUALLY, THE SPREAD OF OVERALL UNCERTAINTY INCREASES…

MOST COMPUTER MODELLING OF NATURAL PHENOMENA HAS ADOPTED A STRATEGY OF SIMPLIFICATION TO MAKE MATTERS TRACTABLE: THOSE DETAILS THAT ARE THOUGHT TO MATTER MOST ARE REPRESENTED AS ACCURATELY AS POSSIBLE, AND OTHER DETAILS, NOT CONSIDERED IMPORTANT, ARE ABRIDGED OR OMITTED. KNOWING, HOWEVER, WHICH DETAILS MATTER MOST CAN BE TRICKY: MODELS ARE PRONE TO BE INCOMPLETE, SOMETIMES LEAVING OUT DETAILS THAT COULD MATTER UNDER CERTAIN CONDITIONS. PARAMETRIC OR EVEN STRUCTURAL

UNCERTAINTIES REMAIN IMPLICIT, SO THAT NO MATTER HOW DETAILED THE MODEL THAT IS CREATED, COMPLETE CONFIDENCE CANNOT BE INVESTED IN ITS PREDICTIONS ABOUT THE BEHAVIOR OF THE REAL SYSTEM... [EMPHASES by author]."[27]

Astute readers no doubt will recognize that the cautions and disclaimers regarding the modeling of Earth's volcanic activity acknowledged by Sparks and Aspinall apply equally to the modeling of Earth's climate—only more so, as the modeling of Earth's long-term volcanic activity is just one aspect of the modeling of Earth's long-term climate. Consequently, the modeling of Earth's long-term volcanic activity must be regarded as an essential partial modeling of Earth's long-term climate. Suffice to say, the heroic modeling challenges presented by such complex phenomena can't be overcome in one fell swoop, and any attempt to model them necessarily must begin with simplified partial models. However, it's important to recognize that such simplification, though necessary, is only the first step down what can be a long hard scientific road which can be traveled for centuries with no end in sight.

The Coupling of Earth's Climate to Solar Activity

Taken together, Earth's climate and solar histories indicate why there would have been ice on the Delaware river near Philadelphia in winter months during the American Revolution and why 1816 was a "year without a summer[99]" during the Little Ice Age—a time in which people in the Northern hemisphere starved when their crops failed due to the extremely cold weather. Clearly, these phenomena were not caused by "anthropogenic activities" or "greenhouse gases," as anthropogenically generated carbon dioxide was *de minimis* at the

---

[27] R. S. J. Sparks and W. P. Aspinall, op cit

time as is evident from the graph in Figure 1. Instead, they were caused by natural astrophysical and geophysical phenomena.

While the "year without a summer" is often attributed to <u>the eruption of Mount Tambora</u>[100] in Indonesia—the largest volcanic eruption in modern geologic history—and the likelihood that it affected Earth's <u>solar climatic forcing</u>[101], the <u>Dalton Minimum</u>[102] in solar activity illustrated in Figure 3 was at least partially responsible—how much of a role it played won't be known until better simulations of Earth's climate have been developed. Attribution of the "year without a summer" to the Tambora volcanic eruption alone simply demonstrates an unwillingness to acknowledge the obvious, namely: that the "year without a summer" coincided with a dramatic reduction in solar activity which undoubtedly had a significant effect on Earth's solar climatic forcing.

As shown in <u>Figure 3</u>[103], the Little Ice Age corresponds to the <u>Maunder Minimum</u>[104] in solar activity (cf. Figure 2), and the "year without a summer" corresponds to the Dalton Minimum. Note also that over the last half of the twentieth century <u>solar activity increased</u>[105] slightly from its average levels during the first half of the twentieth century. Thus, a slight increase in the Earth's average temperature was to be expected in the last half of the twentieth century. The poorly understood phenomena known as <u>elves</u>[106], <u>jets</u>[107], and <u>sprites</u>[108] illustrated in <u>Figure 4</u>[109]—collectively known as "<u>transient luminous events</u>[110]"—electrically couple Earth's troposphere, stratosphere, and mesosphere to Earth's ionosphere and hence to the <u>solar wind</u>[111]—i.e., the widely variable streams of charged particles emanating from chaotic solar mass ejections. These recently discovered and poorly understood transient luminous events have unknown effects on Earth's solar climatic forcing. Thus, scientists have only just begun to understand the coupling of Earth's climate to solar activity.

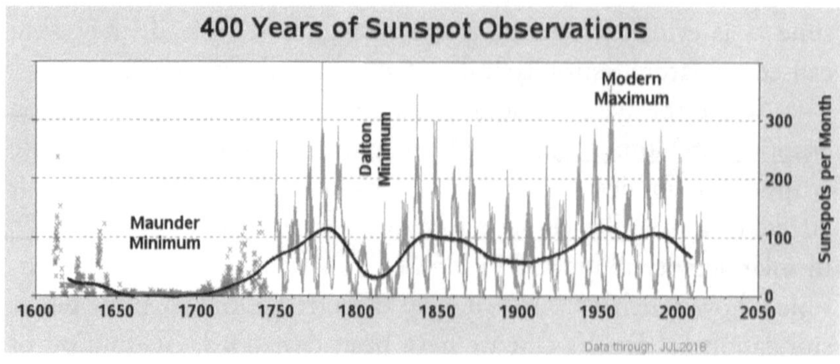

Figure 3

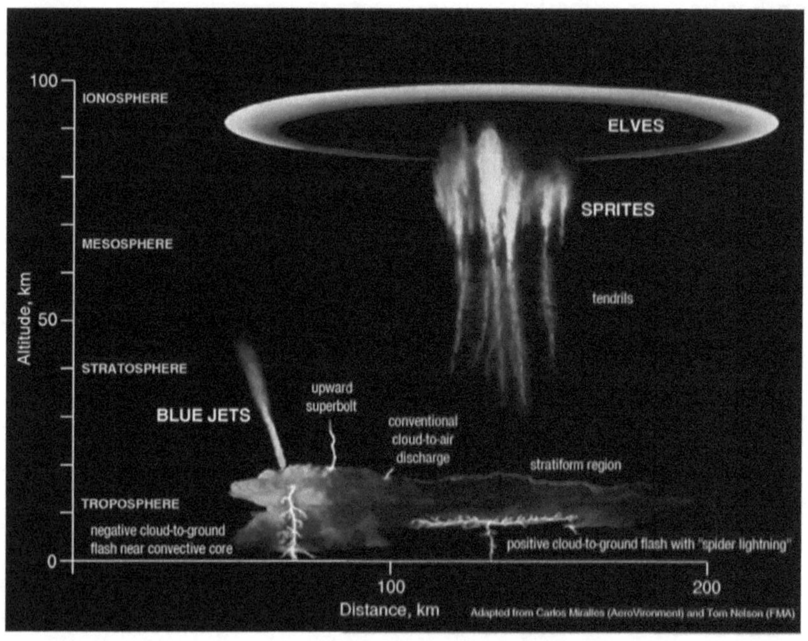

Figure 4

Remarkably, as is evident from the following <u>excerpts</u>[112], the physics of how solar activity effects Earth's climate has only recently attracted the attention of climatologists.

"WHILE THE PHYSICS OF HOW SOLAR ACTIVITY COULD AFFECT CLOUDS REMAINED OBSCURE, IT WAS NOW [in the 1990s] *UNDENIABLE THAT POSSIBLE MECHANISMS COULD EXIST. AND WHILE THE DATA WERE NOISY, A GROWING VARIETY OF EVIDENCE, SOME OF IT GOING BACK THOUSANDS OF YEARS, SHOWED CREDIBLE CORRELATIONS BETWEEN SOLAR ACTIVITY AND ONE OR ANOTHER FEATURE OF THE* [Earth's] *CLIMATE. WHATEVER THE EXACT FORM SOLAR INFLUENCES TOOK, MOST SCIENTISTS WERE COMING TO ACCEPT THAT THE CLIMATE SYSTEM WAS SO UNSTEADY THAT MANY KINDS OF MINOR EXTERNAL CHANGE COULD TRIGGER A SHIFT. IT MIGHT NOT BE NECESSARY TO INVOKE EXOTIC COSMIC RAY MECHANISMS, FOR THE SYSTEM MIGHT BE SENSITIVE EVEN TO THE TINY VARIATIONS IN THE SUN'S TOTAL OUTPUT OF ENERGY, THE* [so-called] *SOLAR CONSTANT. THE BALANCE OF SCIENTIFIC OPINION TILTED. MANY EXPERTS NOW THOUGHT THERE WAS INDEED A SOLAR-CLIMATE CONNECTION.*

When a 1999 study reported evidence that the Sun's magnetic field had strengthened greatly since the 1880s, it brought still more attention to the key question: was increased solar activity the main cause of the rise of average global temperature over that period? As the 21st century began, most experts thought it likely that the Sun had driven at least part of the previous century's warming. MOST CONVINCINGLY, THE WARMING FROM THE

1880s TO THE 1940s HAD COME WHEN SOLAR ACTIVITY HAD DEFINITELY BEEN RISING, WHILE THE CARBON DIOXIDE BUILDUP HAD NOT YET BEEN LARGE ENOUGH TO MATTER MUCH. A COOLING DURING THE 1950s AND 1960s FOLLOWED BY THE RESUMPTION OF WARMING ALSO CORRELATED LOOSELY WITH CHANGES IN SOLAR ACTIVITY. HOW FAR THE SOLAR CHANGES HAD INFLUENCED CLIMATE, HOWEVER, REMAINED SPECULATIVE… ONE SENIOR SOLAR PHYSICIST INSISTED, 'WE WILL HAVE TO KNOW A LOT MORE ABOUT THE SUN AND THE TERRESTRIAL ATMOSPHERE BEFORE WE CAN UNDER-STAND THE NATURE OF THE CONTEM-PORARY CHANGES IN CLIMATE.'…

BUT WHAT IF THE PLANET REALLY DID REACT WITH EXTREME SENSITIVITY TO ALMOST IMPERCEPTIBLE CHANGES IN THE RADIATION ARRIVING FROM THE SUN [EMPHASES by author]?"[28]

What indeed? Note the absurdity of the so-called "solar con-stant[113]" assumption. That egregious astronomical misnomer orig-inated in the over-simplified notion that Earth's solar energy input should be nearly constant—which stemmed from the mistaken assumption that the Sun's energy output is constant and the fact that Earth's elliptical orbit has a relatively small eccentricity so its distance from the Sun is nearly constant (in astronomical terms). Given the intrinsic complexities of its internal magnetohydrodynamics and how little is known about the Sun's production of violent stochastic sun-spots[114], spicules[115], and flares[116], there are few natural phenom-

---

[28] *The Discovery of Global Warming, Changing Sun, "Changing Climate,"* February 2018

ena which are more poorly understood. As it's known that relatively short-lived and small perturbations in Earth's solar energy input can be amplified[29] by the internal <u>nonlinear dynamics</u>[117] of potential stochastic resonances in their <u>couplings to Earth's climate</u>[118], it's evident that Earth's climate may be strongly affected by small perturbations in its solar energy input. And, as acknowledged in the following <u>excerpt</u>[119], these phenomena are fundamental to understanding the Sun's effects on Earth's climate.

> *"Magnetic fields emerging through the solar surface control the Sun's weather, including the dynamics and energetics of its atmosphere and the ejection of high-energy protons and electrons into the solar wind. THESE SOLAR CONDITIONS, IN TURN, IMPACT EARTH'S SPACE WEATHER ENVIRONMENT AND MAGNETOSPHERE. Researchers in the Physics and Astronomy Department at Michigan State University are modeling solar magneto-convection in order to understand how magnetic fields emerge through the Sun's surface, heat the Sun's outer atmosphere, and produce sunspots, spicules, and flares. THIS RESEARCH FURTHERS NASA'S SCIENTIFIC GOALS TO UNDERSTAND THE SUN, ITS VARIATIONS, AND ITS INTERACTIONS WITH EARTH AND THE SOLAR SYSTEM [EMPHASES by author]."*[30]

---

[29] Benzi, op cit

[30] NASA Showcases Agency Science at SC14, *"Understanding the Sun's Weather,"* Overview, 2015

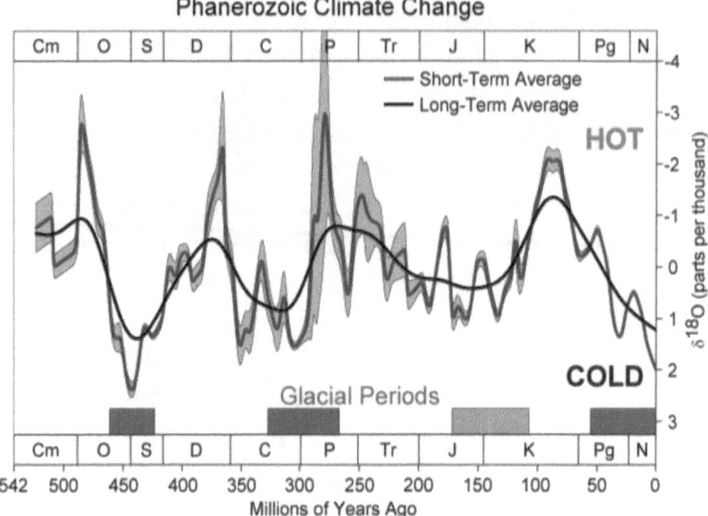

Figure 5

It has long been known that Earth has gone through episodes of extreme climate swings. In fact, as indicated in Figure 5[120], Earth is known to have transitioned through five "Greenhouse Earth" and five "Icehouse Earth" episodes. As it's inconceivable that carbon dioxide alone could account for those realities, the only conceivable conclusion is that there were other phenomena at work which make carbon dioxide's effect on Earth's climate pale into insignificance by comparison. Whatever the causal phenomena were for past "Greenhouse Earth" and "Icehouse Earth" episodes, they certainly were not anthropogenic, and there is little reason to believe that some or all of those phenomena are not still operating today.

Methane: The Elephant in the Room

One doesn't hear or read much about methane[121] in the on-going debates about climate change even though—as noted in the following excerpt[122]—methane is known to be vastly more powerful than carbon dioxide as a heat-trapping gas, and Earth's methane cycles are not well understood.

> *"While carbon dioxide is typically painted as the bad boy of greenhouse gases, METHANE IS ROUGHLY 30 TIMES MORE POTENT AS A HEAT-TRAPPING GAS. New research in the journal Nature indicates that for each degree that Earth's temperature rises, the amount of methane entering the atmosphere from microorganisms dwelling in lake sediment and freshwater wetlands — the primary sources of the gas — will increase several times [EMPHASIS by author]."*[31]

Moreover, that observation says nothing about the potential release of methane currently trapped in <u>methane clathrate</u>[123] crystals embedded in the Earth's oceanic sediments. In "<u>letters to nature</u>[124]," the authors of the following excerpt go so far as to say:

> *"GAS HYDRATES IN OCEANIC SEDIMENTS MAY IN FACT COMPRISE THE EARTH'S LARGEST FOSSIL-FUEL RESERVOIR... OUR RESULTS INDICATE THE PRESENCE OF SUBSTANTIAL QUANTITIES OF METH-ANE (~15 GT OF CARBON) STORED AS SOLID GAS HYDRATE* [in a single reservoir in the Blake plateau off the Southeast Coast of the US], *WITH AN EQUIVALENT OR GREATER AMOUNT OCCURRING AS BUBBLES OF GAS IN THE SEDIMENTS BELOW THE HYDRATE ZONE [EMPHASIS by author]."*[32]

---

[31] ScienceDaily, *"A more potent greenhouse gas than carbon dioxide, methane emissions will leap as Earth warms,"* Princeton University, March 27, 2014

[32] **letters to nature**, *"Direct measurement of in situ methane quantities in a large gas-hydrate reservoir,"* Gerald R. Dickens, et al, 3 December 1996

Earth's oceans are estimated to contain <u>thousands of giga-tons</u>[125], or teratons, of methane in the form of methane clathrate crystals. Interestingly, global cooling can lower Earth's sea levels—thereby reducing the pressure on methane clathrate crystals in the ocean floor causing them to melt and release their methane into the atmosphere. Additionally, global warming can increase the temperature of Earth's oceans—thereby increasing the temperature of the methane clathrate crystals in the ocean floor causing them to melt and release their methane into the atmosphere.

Moreover, Earth's sea levels have varied <u>by as much as four hundred (400) feet</u>[126] and Earth's ocean temperatures have varied widely during episodes of global cooling and warming. As significant decreases in sea levels or increases in ocean temperatures would trigger the release of huge amounts of methane into Earth's atmosphere from the methane clathrate crystals embedded in Earth's oceanic sediments—with <u>potentially explosive effects</u>[127] on Earth's climate, it's clear that any reliable long-term forecast of Earth's climate necessarily must fully account for changes in sea levels and ocean temperatures, as well as their concomitant effects on Earth's atmospheric methane content.

CONCLUSION

Forecasting Earth's Climate: One of Humanity's Greatest Scientific Challenges

For hundreds of years, many of the world's greatest mathematicians and physicists have grappled with the problem of modeling and understanding the physical universe, and there quests have made remarkable progress, but those quests are far from complete. For example, physics has only recently reached a point in its development where a theory of everything[128] is even conceivable. Why then is it so difficult for climatologists to admit they've only just begun their quest to model and understand Earth's climate?

Given climatologists' unabashed acknowledgement that *"A GROWING VARIETY OF EVIDENCE, SOME OF IT GOING BACK THOUSANDS OF YEARS, SHOWED CREDIBLE CORRELATIONS BETWEEN SOLAR ACTIVITY AND ONE OR ANOTHER FEATURE OF THE* [Earth's] *CLIMATE,"*[33] one might expect that climatologists would be more than happy to regard the forecasting of Earth's climate as one of humanity's greatest scientific challenges. And, faced with such a challenge, one might expect some humility from climatologists attempting to make long-term forecasts of Earth's climate. Instead, we are treated to dire forecasts of global disasters predicated on primitive computer models—the reliability of which hasn't been confirmed by unequivocal and verifiable scien-

---

[33] *The Discovery of Global Warming,* op cit

tific observations of predicted phenomena which can't be otherwise explained.

As noted by Sparks and Aspinall, the only means for determining the viability—not to mention the reliability—of computer models is to demonstrate accurate simulations of past phenomena. To quote. Sparks and Aspinall:

> *"INDEED, A TEST FOR THE VIABILITY OF NUMERICAL MODELS SHOULD BE THAT THEIR OUTPUTS MIMIC CLOSELY BOTH THE STATISTICAL AND TEMPORAL PROPERTIES OF THE NATURAL* [i.e., historical] *DATA… A MAJOR CHALLENGE IS TO ENSURE THAT SUCH ENSEMBLE FORE-CASTING ENCOMPASSES ALL THE KEY FACTORS INVOLVED IN THE NATURAL PROCESSES. In particular, the existence of epis-temic* [i.e., known] *uncertainty has to be recognized and assimilated into the procedure, which will entail a suitable suite of alternative models being interro-gated jointly. WHILE, IN PRACTICAL TERMS, THIS WOULD BE A NONTRIVIAL UNDER-TAKING, IT WOULD PROVIDE A PROPER RATIONAL BASIS FOR EVALUATING THE VALUE OF SUCH FORECASTS… IT IS REL-ATIVELY EASY FOR A SKILLED MODELER TO MAKE A MODEL WITH LARGE NUM-BERS OF PARAMETERS FIT THE DATA: THE PROCESS IS ONE OF HINDCASTING, RATHER THAN FORECASTING… WHERE SYSTEMS ARE CLOSE TO THRESH-OLDS FOR INSTABILITY, TECHNICALLY KNOWN AS CUSPS IN CATASTROPHE THEORY, THEY CAN BECOME INHER-ENTLY UNPREDICTABLE. WORRYINGLY, COMPLEX NUMERICAL MODELS ARE*

UNLIKELY TO CAPTURE THIS KIND OF BEHAVIOR EASILY BECAUSE OF THE LARGE EPISTEMIC UNCERTAINTIES IN SOME CRITICAL PARAMETERS [EMPHA-SES by author]."[34]

Note the comment that: *"IT IS RELATIVELY EASY FOR A SKILLED MODELER TO MAKE A MODEL WITH LARGE NUMBERS OF PARAMETERS FIT THE DATA: THE PROCESS IS ONE OF HINDCASTING, RATHER THAN FORECASTING."* That kind of curve fitting is, of course, much easier to accomplish over short time periods. Thus, an ability to simulate short-term climate changes proves nothing about an ability to simulate longer-term climate changes. A short-term model could be simulating nothing more than a short-term perturbation in a longer-term phenomenon. Thus, for example, simulating a short-term rise in global temperature could be the equivalent of simulating a short-term rise in a lava dome which, as noted in the following <u>excerpt</u>[129], proves nothing about the long-term activity of a volcano.

> *"Lava domes evolve unpredictably, due to nonlinear dynamics... The average rate of dome growth may be used as a rough indicator of magma supply, but it shows no systematic relationship to the timing or characteristics of lava dome explosions* [i.e. volcanic eruptions]."

Climatologists' understanding of Earth's climate will fall short of what is required for reliable long-term forecasts until the dynamics of all the major phenomena forcing Earth's climate can be accurately simulated. These phenomena include, but are not limited to: stochastic perturbations in Earth's <u>albedo</u>[130], carbon cycle, magnetic field, methane cycle, orbital dynamics, solar energy input, and the coupling of its solar energy input to Earth's climate—while taking

---

[34] R. S. J. Sparks and W. P. Aspinall, ibid

into full account the internal nonlinear dynamics of potential stochastic resonances in the interactions between them.[35]

If such simulations are not yet feasible, a good starting point would be to back up a step and employ a discipline <u>Peter Drucker</u>[131] described in <u>*LANDMARKS OF TOMORROW*</u>[132] as "organizing our ignorance"—articulated as follows.

> *"Today we increasingly believe that there is a conscious discipline—already learnable though perhaps not yet teachable—for the imaginative leap into the unknown. We are developing rigorous methods for creative perception. Unlike the science of yesterday, it is not based on organizing our knowledge. It is based on organizing our ignorance."*[36]

The discipline Drucker describes begins with compilation of a list of fundamental questions which must be satisfactorily answered as a prerequisite to producing an authoritative understanding of any complex phenomenon. The following questions—in no particular order—are offered as a potential starting point for a list of questions which must be satisfactorily answered as a prerequisite to producing an authoritative understanding of Earth's constantly changing climate.

- Do current climate models accurately simulate Earth's atmospheric carbon dioxide history over thousands and millions of years?
- Do current climate models accurately simulate Earth's atmospheric methane history over thousands and millions of years?
- Do current climate models accurately simulate Earth's carbon and methane cycles and possible stochastic interactions between them over thousands and millions of years?

---

[35] Benzi, ibid

[36] *LANDMARKS OF TOMORROW*, Peter F. Drucker, Transaction Publishers, 1996

- Do current climate models accurately simulate the respective levels of carbon dioxide and methane in Earth's atmosphere before, during, and after "Greenhouse Earth" and "Icehouse Earth" episodes?
- What natural phenomena precipitated "Greenhouse Earth" and "Icehouse Earth" episodes?
- Is there any reason to believe those phenomena don't still effect Earth's climate as they did in the past?
- If those natural phenomena continue to effect Earth's climate as before, do current climate models accurately simulate Earth's climate before, during, and after "Greenhouse Earth" and "Icehouse Earth" episodes?
- If those natural phenomena don't continue to effect Earth's climate as before, what has changed to eliminate them?
- Do current climate models accurately simulate stochastic perturbations in Earth's albedo, carbon cycle, magnetic field, methane cycle, orbital dynamics, solar energy input, and the coupling of its solar energy input to Earth's climate—as well as the internal nonlinear dynamics of potential stochastic resonances in the interactions between them?

As the only means of determining the viability of long-term climate-forecasting models is to demonstrate accurate simulations of Earth's long-term climate changes, unless and until climatologists can provide satisfactory answers to questions like these, intellectual honesty and integrity would suggest they temper their claims of a capability to make reliable long-term forecasts of Earth's climate. Until then, claims of an ability to reliably forecast Earth's climate for hundreds of years or longer periods of time are simply preposterous.

As geophysicists and volcanologists—such as Sparks and Aspinall—have no trouble admitting they still face daunting challenges in making reliable short-term forecasts of Earth's volcanic activity, it's somewhat mystifying why climatologists feel compelled to make claims of an ability to make reliable long-term forecasts of Earth's climate—which challenge is at least as difficult, if not more so, than the challenge of making reliable short-term forecasts of Earth's volcanic activity.

The Tragedy of It All

Evidently, some climatologists are embarrassed by the fact they haven't yet been able to model and accurately simulate some of the phenomena known to be forcing Earth's climate changes, and consequently, they appear to have adopted a strategy of claiming any effects on Earth's climate by such phenomena are insignificant by comparison to the effects of atmospheric carbon dioxide. Unfortunately, there is no scientific evidence on which to base such a claim. Unless and until the dynamics of all the phenomena known to be forcing Earth's climate changes can be modeled and accurately simulated—while taking into account the internal nonlinear dynamics of potential stochastic resonances in the interactions between them—their individual and collective effects on Earth's climate can't be known. Thus, it's absurd to claim the effects of some of those phenomena are insignificant compared to others without first being able to model and accurately simulate them to ascertain what individual and collective effects they really have on Earth's climate.

Given that heroic scientific challenges—like those associated with simulating Earth's volcanic activity and climate changes—can't be overcome in one fell swoop, scientists have no alternative but to begin with simple partial models. Of course, such models are primitive first steps down what can be a long hard scientific road with no end in sight. However, there's no need for embarrassment in admitting there's no end of the road in sight when dealing with such heroic scientific challenges. On the other hand, buying into the delusional proposition that reliable long-term forecasts can be made of Earth's climate in the absence of a demonstrated ability to accurately simulate some of the most powerful phenomena forcing Earth's climate changes is counterproductive, shameful, and unworthy of anyone claiming to be a scientist.

Clearly, climatologists didn't deliberately set out to deceive the public, but the contrast between how climatologists and volcanologists—often from the same research institutions—characterize the reliability of their forecasts demands an explanation. The only satisfactory explanation seems to be that climatologists allowed themselves to be caught up in a scientific fraud perpetrated by a vile combination of AGW demagogues; demented billionaires; foreign governments; mercenary businesses; and Neo-Marxist foundations, institutions, and revolutionaries. In other words, climatologists likely were overwhelmed by the same forces which overwhelmed the publishers of weekly and monthly periodicals—combined with the irresistible appeal of increased government funding, personal celebrity and self-importance.

With climatologists conspicuously unable to prove their AGW hypothesis, AGW demagogues are now desperately trying to change the subject. By branding nonbelievers as "climate deniers" and "science deniers," AGW demagogues are falsely claiming nonbelievers deny the well-established fact that Earth's climate changes over time. Nothing could be further from the truth. Such pathetic attempts to change the subject are nothing more, nor less, than the classic smokescreen strategy employed by demagogues whenever they're caught in lies of their own making. The real "climate deniers" are those who are so ignorant of our physical world they don't know the only thing constant about Earth's climate is that it is now, and always has been, constantly changing—and changing so chaotically the changes can be accurately described only as stochastic.

Tragically, humankind display an innate inability to behave rationally when panicked. Unfortunately, in this regard, humankind are no different than other species of the Mammalia class of animals. Sadly, there is no cure for this herd-like behavior. Moreover, every demagogue has at least an intuitive understanding of this weakness and knows the vast majority of people are so gullible and ignorant they can be stampeded like herds of cattle if only they are sufficiently frightened. Consequently, the advent of inexpensive, stealthy, and universally available mass media and their perfection as instruments for disseminating agitprop enables demagogues to frighten and stampede the vast majority of people into blindly participating in conspiracies motivated solely by pecuniary and political power lust.

Thus, humanity is now confronted by an entirely new kind of threat, namely: the ease with which relatively small numbers of self-appointed, self-aggrandizing, self-perpetuating, self-righteous, self-serving, and supercilious so-called "elites"—aided and abetted by demagogues; demented billionaires; foreign governments; mercenary businesses; and Neo-Marxist foundations, institutions, and revolutionaries—are able to stampede the masses into pursuing the political and profit agendas of the so-called "elites" irrespective of how demented, ill-conceived, or counterproductive those agendas may be to human freedoms, liberties, social progress, and technological advancement.

The irony of the science fiction in the <u>Planet of the Apes</u>[133] is that what is portrayed as a potential future human evolution instead is a metaphor for past human evolution. Whether intentional or not by its authors, that metaphorical mirror of past human evolution should be a highly instructive warning to humankind not to fall back on their innate animalism when confronted by fear and panic incited by demagoguery and exacerbated by massive human gullibility and ignorance. In this regard, the current climate-change hysteria should serve as a prime example of precisely the kind of human behavior to be avoided if at all possible.

Rather than continuing to further inflame climate-change hysteria, isn't it about time to admit the obvious reality that Earth's climate is, and always has been, constantly changing far beyond any feeble human capability to stop it and acknowledge that humanity should focus on accommodating what it can't control? To do otherwise would be just one more pathetic demonstration of human gullibility and ignorance.

Rather than exploiting—as Al Gore and other demagogues and unscrupulous profiteers have been doing—ways to profit from pandemic global climate-change hysteria, why not focus on all of the wonderful innovations and whole new industries which will be needed to successfully cope with the natural phenomenon of Earth's constantly changing climate and what marvelous new opportunities they present for humanity? Successfully coping with Earth's natural climate changes truly is one of humanity's greatest challenges—on a par with surviving asteroid collisions, supervolcano eruptions, and finding a way to escape planet Earth before the Sun becomes a Red Giant[134]. Tragically, humankind are almost certainly doomed to eventual extinction unless they can learn to focus their energies, resources, and talents on developing real solutions to real scientific challenges—while avoiding their "Chicken Little[135]" tendency to be stampeded into counterproductive wheel-spinning exercises by the combined forces of political demagogues; demented billionaires; foreign governments; mercenary businesses; and Neo-Marxist foundations, institutions, and revolutionaries.

ENDNOTES

1. https://en.wikipedia.org/wiki/The_Boy_Who_Cried_Wolf
2. https://www.wsj.com/articles/how-the-exxon-case-unraveled-1472598472
3. http://www.ipcc.ch/
4. https://en.wikipedia.org/wiki/Milutin_Milankovi%C4%87
5. https://en.wikipedia.org/wiki/Agitprop
6. https://en.wikipedia.org/wiki/Fourth_Estate
7. https://en.wikipedia.org/wiki/Thomas_Carlyle
8. https://en.wikipedia.org/wiki/Ferdinand_Lundberg
9. https://en.wikipedia.org/wiki/Carbon_cycle
10. http://adsabs.harvard.edu/full/1993AREPS..21..407W
11. https://en.wikipedia.org/wiki/Milankovitch_cycles
12. https://en.wikipedia.org/wiki/Solar_cycle
13. https://oceanservice.noaa.gov/facts/ninonina.html
14. https://i.redd.it/8q8yevlvhwf21.jpg
15. https://en.wikipedia.org/wiki/Communist_International
16. https://www.youtube.com/watch?v=Cf2nqmQIfxc
17. https://en.wikipedia.org/wiki/Neo-Marxism
18. https://www.amazon.com/Draining-Swamp-Survive-Socio communist-Societal-ebook/dp/B078GZBW9B
19. https://www.wsj.com/articles/100-years-of-communismand-100-million-dead-1510011810
20. https://en.wikipedia.org/wiki/The_War_of_the_Worlds_(radio_drama)
21. https://en.wikipedia.org/wiki/Illusory_truth_effect
22. https://www.sheffield.ac.uk/psychology/staff/academic/tom-stafford

23. http://www.bbc.com/future/story/20161026-how-liars-create-the-illusion-of-truth?ocid=fbfut
24. https://en.wikipedia.org/wiki/Joseph_Goebbels
25. https://en.wikipedia.org/wiki/Karl_Marx
26. https://en.wikipedia.org/wiki/Vladimir_Lenin
27. https://en.wikipedia.org/wiki/Adolf_Hitler
28. https://en.wikipedia.org/wiki/Saul_Alinsky
29. https://www.econlib.org/library/Mises/msS.html?chapter_num=5#book-reader
30. https://en.wikipedia.org/wiki/Ludwig_von_Mises
31. https://en.wikipedia.org/wiki/William_James
32. https://www.enotes.com/topics/pragmatism/summary
33. https://en.wikipedia.org/wiki/Elizabeth_Nickson
34. http://www.freerepublic.com/focus/f-news/885796/posts
35. https://en.wikipedia.org/wiki/Dwight_D._Eisenhower
36. http://www.americanrhetoric.com/speeches/dwightdeisenhower-farewell.html
37. https://en.wikipedia.org/wiki/Peer_review
38. https://en.wikipedia.org/wiki/Principal_investigator
39. https://en.wikipedia.org/wiki/Regulatory_capture
40. https://en.wikipedia.org/wiki/George_Stigler
41. https://en.wikipedia.org/wiki/Saul_Alinsky
42. https://en.wikiquote.org/wiki/Rahm_Emanuel
43. https://en.wikipedia.org/wiki/James_Hansen
44. https://en.wikipedia.org/wiki/Al_Gore
45. https://www.scientificamerican.com
46. https://www.scientificamerican.com/store/books/return-to-reason-the-science-of-thought/
47. https://en.wikipedia.org/wiki/Katharine_Hayhoe
48. https://en.wikipedia.org/wiki/Dysrationalia
49. https://www.scientificamerican.com/store/books/climate-change-planet-under-pressure/
50. https://en.wikipedia.org/wiki/Greenhouse_and_icehouse_Earth#Greenhouse_Earth
51. https://en.wikipedia.org/wiki/List_of_parties_to_the_Paris_Agreement

52. https://en.wikipedia.org/wiki/Dark_Ages_(historiography)
53. https://en.wikipedia.org/wiki/John_Dalberg-Acton,_1st_Baron_Acton
54. https://en.wikipedia.org/wiki/Jeff_Bezos
55. https://en.wikipedia.org/wiki/Michael_Bloomberg
56. https://en.wikipedia.org/wiki/Tom_Steyer
57. https://en.wikipedia.org/wiki/Donald_Sussman
58. https://en.wikipedia.org/wiki/George_Soros
59. https://en.wikipedia.org/wiki/Judas_goat
60. https://en.wikipedia.org/wiki/Donald_Trump
61. https://en.wikipedia.org/wiki/Flash_mob
62. https://en.wikipedia.org/wiki/Antifa_(United_States)
63. https://en.wikipedia.org/wiki/Abraham_Lincoln
64. http://www.abrahamlincolnonline.org/lincoln/speeches/gettysburg.htm
65. https://en.wikipedia.org/wiki/Sturmabteilung
66. https://www.britannica.com/place/Third-Reich
67. https://en.wikipedia.org/wiki/Henning_Webb_Prentis_Jr.
68. http://ergo-sum.net/literature/CultOfCompetency.pdf
69. https://www.quora.com/profile/Kenneth-Archer
70. https://www.quora.com/How-hard-would-it-be-to-conquer-the-U-S
71. https://en.wikipedia.org/wiki/%20Thomas_Jefferson
72. https://www.sciencebuddies.org/science-fair-projects/science-fair/steps-of-the-scientific-method
73. https://en.wikipedia.org/wiki/Albert_Einstein
74. https://en.wikipedia.org/wiki/General_relativity
75. https://en.wikipedia.org/wiki/Tests_of_general_relativity
76. Medieval Warm Period and a Little Ice Age
77. https://www.youtube.com/watch?v=sTPwq7ofySk
78. https://en.wikipedia.org/wiki/Asteroid
79. https://en.wikipedia.org/wiki/Comet
80. https://en.wikipedia.org/wiki/Planet
81. https://en.wikipedia.org/wiki/Fundamental_ephemeris
82. https://en.wikipedia.org/wiki/Perturbation_(astronomy)

83. https://www.nature.com/scitable/knowledge/library/milankovitch-cycles-paleoclimatic-change-and-hominin-evolution-68244581

84. http://www.apricus.com/html/solar_collector_insolation.htm#.W7GGOXtKjRY

85. https://en.wikipedia.org/wiki/Beat_(acoustics)

86. https://en.wikipedia.org/wiki/Milankovitch_cycles#Orbital_inclination

87. http://adsabs.harvard.edu/full/1993AREPS..21..407W

88. https://en.wikipedia.org/wiki/Apsidal_precession

89. https://en.wikipedia.org/wiki/Orbital_inclination

90. https://www.nonlin-processes-geophys.net/17/431/2010/npg-17-431-2010.pdf

91. https://www.wsj.com/articles/the-climate-snow-job-1453664732

92. https://en.wikipedia.org/wiki/Patrick_Michaels

93. https://www.nature.com/scientificamerican/journal/v315/n4/box/scientificamerican1016-68_BX2.html

94. https://en.wikipedia.org/wiki/Supervolcano

95. https://phys.org/news/2018-02-fluctuations-earth-magnetic-field.html

96. https://en.wikipedia.org/wiki/Geomagnetic_reversal

97. https://www.nature.com/news/earth-s-lost-history-of-planet-altering-eruptions-revealed-1.21630

98. qhttp://dutiosc.twi.tudelft.nl/~risk/extrafiles/EJcourse/Sheets/SparksAspinall_VolcanicActivity.pdf

99. https://en.wikipedia.org/wiki/Year_Without_a_Summer

100. https://en.wikipedia.org/wiki/1815_eruption_of_Mount_Tambora

101. https://en.wikipedia.org/wiki/Radiative_forcing

102. https://en.wikipedia.org/wiki/Dalton_Minimum

103. http://robslink.com/SAS/democd24/sunspot.htm

104. https://en.wikipedia.org/wiki/Maunder_Minimum

105. https://www.space.com/2942-sun-activity-increased-century-study-confirms.html

106. https://weather.com/news/news/transient-luminous-events-mysteries-sky-20130731

107. https://www.amusingplanet.com/2016/09/the-elusive-gigantic-jets-of-lightning.html
108. https://en.wikipedia.org/wiki/Sprite_(lightning)
109. https://upload.wikimedia.org/wikipedia/commons/9/93/Lightning_sprites.jpg
110. https://en.wikipedia.org/wiki/Upper-atmospheric_lightning
111. https://en.wikipedia.org/wiki/Solar_wind
112. https://history.aip.org/climate/solar.htm
113. https://books.google.com/books?id=xjnvCAAAQBA-J&pg=PA182&lpg=PA182&dq=thermo+hydrody-namics+of+the+sun&source=bl&ots=bI4fppd9If&sig=vhVe6Pp-3Paw5DaWFybPE5C7gOhA&hl=en&sa=X&ved=2ahUKEw-jbwp_gt-jdAhUGd6wKHRQqAwQQ6AEwDHoECAAQA-Q#v=onepage&q=thermo%20hydrodynamics%20of%20the%20sun&f=false
114. https://en.wikipedia.org/wiki/Sunspot
115. https://en.wikipedia.org/wiki/Spicule_(solar_physics)
116. https://en.wikipedia.org/wiki/Solar_flare
117. https://en.wikipedia.org/wiki/Chaos_theory
118. https://hallingblog.com/tag/anthropomorphic-global-warming/
119. https://www.nas.nasa.gov/SC14/demos/demo6.html
120. https://en.wikipedia.org/wiki/Geologic_temperature_record#/media/File:Phanerozoic_Climate_Change.png
121. https://www.giss.nasa.gov/research/features/200409_methane/
122. https://www.sciencedaily.com/releases/2014/03/140327111724.htm
123. https://en.wikipedia.org/wiki/Methane_clathrate
124. https://deepblue.lib.umich.edu/bitstream/han-dle/2027.42/62828/385426a0.pdf;jsessionid=3F5C4CCFD7F-FAF298B32C394DB21963F?sequence=1
125. https://worldoceanreview.com/en/wor-1/ocean-chemistry/climate-change-and-methane-hydrates/
126. https://ocean.si.edu/through-time/ancient-seas/sea-level-rise#section_402
127. https://en.wikipedia.org/wiki/Clathrate_gun_hypothesis
128. https://en.wikipedia.org/wiki/Theory_of_everything

129. https://en.wikipedia.org/wiki/Lava_dome
130. https://en.wikipedia.org/wiki/Albedo
131. https://en.wikipedia.org/wiki/Peter_Drucker
132. https://www.passeidireto.com/arquivo/51292579/drucker-1957-1/8
133. https://en.wikipedia.org/wiki/Planet_of_the_Apes
134. https://www.universetoday.com/12648/will-earth-survive-when-the-sun-becomes-a-red-giant/
135. https://en.wikipedia.org/wiki/Chicken_Little

ABOUT THE AUTHOR

David L. R. (Dave) Stein,

Serial Entrepreneur (retired)

Dave Stein lives in Austin, Texas. He is a computer-industry and start-up veteran. Over the last fifty years Dave has been a director of ten high-technology start-up companies, including four with combined revenues of more than $1 billion that he cofounded. His track record includes twenty years in the computer industry in a variety of engineering, sales, marketing, and general management positions at Control Data, IBM, Scientific Data Systems, Univac, Systems Engineering Laboratories, and Harris Corporation. In 1979, he cofounded Gartner Inc. where he managed all operations from day one, grew recurring revenues to $40 million in six years, and established Gartner's franchise as the global leader in strategic IT consulting services. In 1985, he cofounded a $250 million venture-capital partnership in Southern California. From 1992 to 2004, he was a management consultant in San Jose. From 2004 to 2016, he was cofounder of two start-up software companies. He received a BS degree with distinction in Mathematics and Physics and did graduate work in Mathematics at the University of Minnesota Institute of Technology.

www.ingramcontent.com/pod-product-compliance
Lightning Source LLC
Chambersburg PA
CBHW031153250726

48655CB00002B/950